ENCYCLOPEDIA OF EARTH SCIENCE

ENCYCLOPEDIA OF EARTH SCIENCE

Edited by

Dr. Kailash Malode

Delve PUBLISHING

www.delvepublishing.com

Encyclopedia of Earth Science

Dr. Kailash Malode

Delve Publishing
224 Shoreacres Road
Burlington, ON L7L 2H2
Canada
www.delvepublishing.com

Email: orders@arclereducation.com

ISBN: 978-1-77469-390-2 (Hardcover)

ABOUT THE EDITOR

Dr. Kailash Malode (1980) is presently serving as Assistant Professor, Department of Soil Science And Agriculture Chemistry, college of Agriculture, Parul University, Vadodara, Gujarat. He obtained his B. Sc.(Ag.) in 2006 from Marathwada Agriculture University, Parbhani. M. Sc. (Ag.)2009 in Dr. Panjabrao Deshmukh Krishi Vidyapeeth, Akola and Ph. D. Soil Science And Agriculture Chemistry in 2014 from Vasantrao Naik Marathwada Krishi Vidhyapeeth, Parbhani. He started his career as Assistant professor in 2014. His field of specialization is soil fertility, soil carbon sequestration and organic carbon management. He has edited 06 books at international level, published 19 research papers in National and international journals of reputed. He has participated in many National, State Seminars and symposiums. He has written 02 practical manuals for the undergraduate course for the benefit of students, besides several radio talks and on farmers on a farmer training programme. Dr. Kailash has also written 26 popular articles in Shati Bharti, Shati pragati and other reputed magazines and 22 audio talks. He had the Best writer award in 2015 for Krishi dooth popular Magazines of Maharashtra, Young Scientists Award 2021 by Astha foundation and Birsa Agriculture University, Jharkhand. Dr. Kailash got the Best scientist award in soil science 2023 by Rajmata Vijayaraje Sindhya Krishi Vishva Vidhyalaya, Gwalior, and also best poster presentation of same.

TABLE OF CONTENTS

LIST OF FIGURES

LIST OF ABBREVIATIONS

AU	astronomical units
BC	bolometric correction
BWR	boiling water reactors
CCL	convective condensation level
CO_2	carbon dioxide
DAC	diamond anvil cell
DEM	digital elevation data
EEI	earth energy imbalance
EMF	electromagnetic force
EPA	Environmental Protection Agency
ESA	European Space Agency
HVS	human visual system
HWR	heavy-water reactors
LAB	lithosphere-asthenosphere border
LCL	lifting condensation level
LEED	leadership in energy and environmental design
LEWP	line echo wave pattern
MORB	mid-ocean ridge
PSCs	polar stratospheric clouds
PWR	pressurized water reactors
QLCS	quasi-linear convective system
UV	ultraviolet

ABSTRACT

Earth studies, its beginning and structure, makeup, and history, and the idea of natural cycles which have led to its existing forms is known as Earth science. It comprises of a. Crystallography, geochemistry, mineralogy, and petrology which are focused on Earth constituents b. Structural topography and geophysics which are about the structure and setup of the different units whereby the Earth comprises c. Stratigraphy and historical topography which are about the delineation and age connection of Earth resources d. Fossil science which is about early life in relation with Earth materials, and e. Physical topography and geomorphology about surface topographical cycles and the successive landforms. Earth science education has gained significantly over the recent 20 years. The reason is that the mineral abundance of countries is to be commercially tapped for the development of the financial conditions and increase in national prosperity, developing nations should be all-around educated and trained in Earth Sciences. Involving sample collection would also aid the students' understanding of the topics discussed.

PREFACE

Earth science involves the study of the globe's Earth's structure, properties, cycles, and 4.5 billion years of biotic development. Insight on the composition, stratigraphy, and chemical structure of the world's crust assists with finding resources that maintain and promote the nature of life. Earth science examines terrestrial surface and organic resources (water, metals, ground, soil, and stones) just as cultivation. Earth science concentrates on the structure, cycles, and planet properties, alongside almost 5 billion years of biological development. The phrase "earth sciences" incorporates the study of the climate (meteorology), water streaming on or underneath planetary surfaces (hydrology), and the seas and oceans (oceanography). Continuous advancements in studies of the planet give an understanding of evolving conditions, the natural circulation of energy resources, and give different ways to relieve and foreseeing the impacts of topographical catastrophes like seismic tremors, floods, volcanic discharges, and avalanches.

This book discusses the four key divisions of Earth science which are geography, oceanography, meteorology, and astronomy. Geography involves the study of the geosphere, consisting of Earth's stones and minerals. Meteorologists concentrate on the air and how its capacities concerning climate and environment. Oceanography involves research on the global oceans. These waters transport moisture all over the planet and hence have a significant impact on atmospheric warming. Cosmology involves study of everything in the universe beyond Earth's climate. That incorporates objects we can see with unaided eyes, such as the Sun, moon, planets, and stars.

Earth science involves thinking globally and acting locally—to use wise judgment about issues significant in everyday life as people and residents. Individuals who understand the Earth's system can make informed choices regarding where to invest in natural resources. Our lives and development rely on how we comprehend and deal with our planet—Earth processes influence all of us. Climate change impact the accessibility of water resources, whereas earthquakes, volcanic emissions, typhoons, and floods can kill people and cause millions of dollars in property destruction.

Similarly, Earth processes affect every one of us – as individuals, societies, and countries. As we utilize these resources, we affect Earth directly which will thus influence the generations which come later on. To improve environmental stewardship, its key to continue into the future with a sound comprehension of Earth structures. The Earth is a complex mechanism consisting of land, water, and air which support every other living thing roaming the planet.

Earth science knowledge empowers us to think about the world and act locally—to use wise judgment about issues significant in our lives as people and global residents. Individuals who see how Earth structures work can settle on informed choices regarding where to purchase or create a home free from environmental danger. They can discuss and resolve issues encompassing clean water, metropolitan planning and improvement, public safety, worldwide environmental change, and the utilization of natural resources. An educated society, aware of our intricate relations with the planet, perceives the significance of and demands Earth science education on all grade levels—rudimentary, secondary, and adult school. When we underscore the significance of Earth science training, everybody benefits.

A

Ablation

A combination of processes (such as sublimation, fusion or melting, and evaporation) that remove snow or ice from the surface of a glacier or snowfield; also used to represent the amount lost by these processes. (2) melting, evaporation, wind, and avalanches reduce the water equivalent of a snow cover (Yu and Gong, 2012).

Ablation, the inverse of accumulation, refers to all processes that remove snow, ice, or water from a glacier or snowfield in glaciology and meteorology. Melting of snow or ice that runs off the glacier, evaporation, sublimation, calving, or erosive removal of snow by wind are all examples of ablation. Typically, air temperature is the major control of ablation, with precipitation acting as a secondary control. During ablation season in a temperate climate, ablation rates normally average around 2 mm/hr. When solar radiation is the primary driver of snow ablation, the snow surface may acquire distinctive ablation textures such as Suncups and Penitentes (e.g., if air temperatures are low under clear skies).

Abnormal

Deviating from what is normal or usual, typically in a way that is undesirable or worrying.

Abrasion

Abrasion is an erosion process that occurs when the material being transported wears away at a surface over time. It is the process of scuffing, scratching, wearing down, marring, and rubbing away of materials induced by friction. The abrasion intensity is determined by the hardness, concentration, velocity, and mass of the moving particles. Abrasion can happen in four different ways (Witze, 2013). Glaciation grinds boulders scooped up by ice against rock surfaces over time. In river channels, solid items produce abrasive surface contact with the bed and walls. Abrasion is caused by objects transported in waves breaking on coasts. Finally, wind moving sand or small stones on surface rocks can create abrasion.

Absolute

The absolute age of a material on Earth is a measure of how ancient it is in years. Some scientists prefer to refer to it as calendar age since it implies that the age may be plotted on a calendar.

For example, if a sedimentary rock layer is wedged between two layers of volcanic ash, its age is somewhere between the ages of the ash layers. Geologists have built a well-defined timeline of Earth history using a mix of radiometric dating, index fossils, and superposition (Wilkinson et al., 2020).

Relative dating does not provide precise dates; it just determines if one artifact, fossil, or stratigraphic layer is older than another. Absolute dating methods provide more precise origin dates and temporal ranges, such as a year-by-year age range.

Absolute Magnitude

Absolute magnitude (M) is a measure of a celestial object's luminosity on an inverse logarithmic astronomical magnitude scale. The absolute magnitude of an object is defined as the apparent magnitude of the object if viewed from a distance of exactly 10 parsecs (32.6 light-years), with no extinction (or dimming) of its light owing to absorption by interstellar matter and cosmic dust (Wallace-Wells, 2019). By supposing all objects to be at a standard reference distance from the observer, their luminosities can be directly compared on a magnitude scale.

As with all astronomical magnitudes, the absolute magnitude can be defined for multiple wavelength ranges corresponding to specified filter bands or passbands; for stars, the absolute visual magnitude, which employs the visual (V) band of the spectrum, is an often-quoted absolute magnitude (in the UBV photometric system).

Absolute magnitudes are denoted by a capital M, with a subscript signifying the measurement filter band, such as MV for absolute magnitude in the V band (Figure 1).

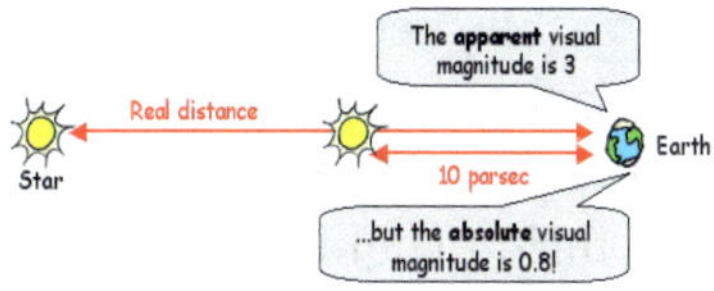

Figure 1. Absolute magnitude.

Source: https://astronomy.swin.edu.au/cosmos/a/Absolute+Magnitude.

The smaller the numerical value of an object's absolute magnitude, the brighter it is. A difference of 5 magnitudes in absolute magnitude between two objects corresponds to a luminosity ratio of 100, while a difference of n magnitudes in absolute magnitude corresponds to a luminosity ratio of 100n/5. In the V filter band, a star with absolute magnitude MV=3.0 would be 100 times brighter than a star with absolute magnitude MV=8.0. The Sun has an absolute magnitude of +4.83 (MV=+4.83). Highly luminous objects can have negative absolute magnitudes: the Milky Way galaxy, for example, has an absolute B magnitude of around 20.8 (Zhuang et al., 2019).

The absolute bolometric magnitude (Mbol) of an object represents its entire luminosity across all wavelengths, rather than in a particular filter band, as stated on a logarithmic magnitude scale. A bolometric correction (BC) is used to convert an absolute magnitude in a specified filter band to an absolute bolometric magnitude.

A distinct definition of absolute magnitude (H) is used for Solar System bodies that shine in reflected light, based on a standard reference distance of one astronomical unit.

Absolute Zero

Absolute zero is the lowest limit of the thermodynamic temperature scale, a state in which the enthalpy and entropy of a cooled ideal gas reach 0 kelvins. Nature's fundamental particles contain very little vibrational motion, preserving only quantum mechanical, zero-point energy-induced particle motion. The theoretical temperature is calculated by extrapolating the ideal gas law; absolute zero is defined by international consensus as 273.15°C (International System of Units), which equals 459.67°F (United States customary units or Imperial units). The zero points of the equivalent

Kelvin and Rankine temperature scales are, by definition, fixed at absolute zero (Toth et al., 2014).

It is commonly thought to be the lowest temperature possible, but it is not the lowest enthalpy state possible, because when cooled, all real substances begin to deviate from the ideal gas as they approach the change of state to liquid, and then to solid; and the sum of the enthalpy of vaporization (gas to liquid) and enthalpy of fusion (liquid to solid) exceeds the ideal gas's change in enthalpy to absolute zero. Matter (solid) at absolute zero is in its ground state, or point of lowest internal energy, according to the quantum-mechanical description (Figure 2).

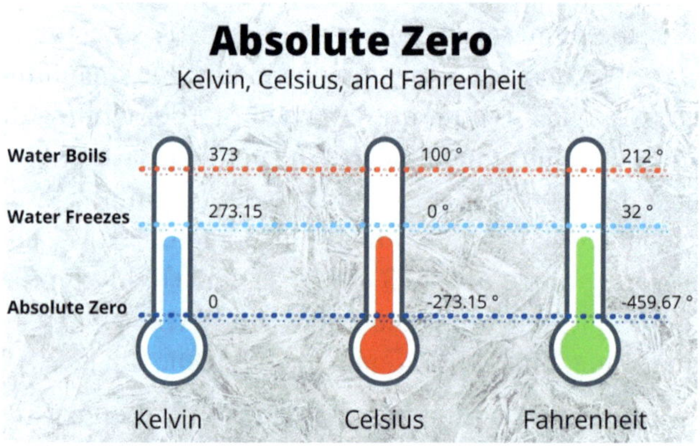

Figure 2. Absolute zero is 0 K, –273.15°C, or –459.67°F.

Source: https://sciencenotes.org/what-is-absolute-zero-temperature-in-kelvin-celsius-and-fahrenheit/.

The laws of thermodynamics indicate that absolute zero cannot be attained solely through thermodynamic means, because the temperature of the substance being cooled approaches the temperature of the cooling agent asymptotically, and a system at absolute zero retains quantum mechanical zero-point energy, the energy of its ground state at absolute zero. The ground state's kinetic energy cannot be removed (Tollefson, 2015).

Scientists and engineers often reach temperatures near absolute zero, where matter exhibits quantum properties such as Bose-Einstein condensate, superconductivity, and superfluidity.

Absorb

To take in, wave motion, absorption is the transfer of a wave's energy to matter when the wave travels through it. The energy of an acoustic, electromagnetic, or other wave is proportional to the square of its amplitude—that is, the maximum displacement or movement of a point on the wave—and the amplitude continuously diminishes as the wave goes through a substance. If only a tiny fraction of the energy is absorbed, the medium is said to be transparent to that particular radiation; nevertheless, if all of the energy is lost, the medium is said to be opaque (Thessen, 2016). To some extent, all known transparent substances absorb. For example, the water seems transparent to sunlight near the surface but gets opaque as one descends.

Substances absorb radiation selectively, that is, they absorb radiation of specific wavelengths. Hard rubber is transparent to infrared and X-rays but opaque to visible light; green glass is transparent to green light but opaque to blue and red; and hard rubber is transparent to green light but opaque to blue and red. Radiation of an undesired wavelength can thus be eliminated from a wave combination by passing it through a suitable medium. Filters are compounds that are designed to absorb a specific wavelength or band of wavelengths.

Absorption Spectrum

The spectrum of electromagnetic radiation that has passed through a medium that absorbed radiation of certain wavelengths. The spectrum is used to measure absorption where various wavelengths of light represent different colors of light absorption spectrum.

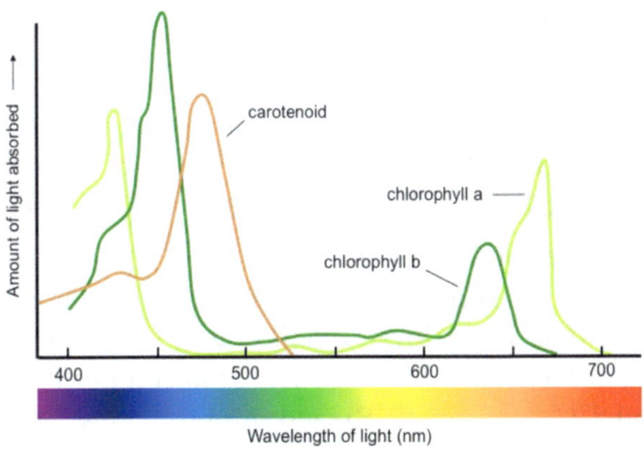

Figure 3. Absorption and action spectra.

Source: https://ib.bioninja.com.au/standard-level/topic-2-molecular-biology/29-photosynthesis/action-spectrum.html.

(Science: chemistry) a graph of the amount of light a substance absorbs, plotted as a function of energy, frequency, or wavelength (Telling et al., 2017).

An electromagnetic spectrum in which a decrease in intensity of radiation at specific wavelengths or ranges of wavelengths characteristic of an absorbing substance is manifested especially as a pattern of dark lines or bands (Figure 3).

Abundant

- Having a plentiful supply (of).
- (of a chemical element or mineral) occurring to an extent specified in relation to other elements or minerals in the Earth's crust or some other specified environment (Showstack, 2008).
- (of an isotope) occurring to an extent specified in relation to other isotopes in a mixture of isotopes.

Abyssal

The abyssal zone is a region of the ocean that is deeper than 2,000 m (6,600 ft) but shallower than 6,000 m. (20,000 feet). The zone is primarily characterized by its extremely uniform environmental conditions, which are reflected in the various living types that inhabit it. The upper border between the abyssal zone and the overlying bathyal zone is easily defined as the depth where the water temperature is 4°C (39°F); this depth ranges between 1,000 and 3,000 m. Ecologists refer to water depths more than 6,000 meters as the hadal realm.

The abyssal realm is Earth's greatest habitat, covering 300,000,000 square kilometers (115,000,000 square miles), or around 60% of the planetary surface and 83% of the area of oceans and seas (Schmitz, 2011).

Abyssal waters form at the air-sea interface in polar locations, most notably Antarctica. The frigid temperature creates marine ice and residual cold brine there. The brine lowers and slowly travels along the bottom toward the Equator due to its high density. Temperatures in the abyss range between 0° and 4°C (32° and 39°F), and salinities range between 34.6 and 35.0 parts per thousand. With each 10-meter increment in-depth, pressure rises by around one atmosphere (about 14.7 pounds per square inch at sea level); hence, abyssal pressures range between 200 and 600 atmospheres. Pressure, on the other hand, poses minimal challenges for abyssal organisms because the pressures within their bodies are the same as those outside of them.

Accessory

A thing which can be added to something else in order to make it more useful, versatile, or attractive. Accessory mineral, any mineral in an igneous rock not essential to the naming of the rock. When it is present in small amounts, as is common, it is called a minor accessory. If the amount is greater or is of special significance, the mineral is called a varietal, or characterizing, accessory, and may give a varietal name to the rock (e.g., the mineral biotite in biotite granite). Accessory minerals characteristically are formed during the solidification of the rocks from the magma; in contrast are secondary minerals, which form at a later time through processes such as weathering by hydrothermal alteration (Schmitz, 2012). Common minor accessory minerals include topaz, zircon, corundum, fluorite, garnet, monazite, rutile, magnetite, ilmenite, allanite, and tourmaline. Typical varietal accessories include biotite, muscovite, amphibole, pyroxene, and olivine.

Accompanying

Provided or occurring at the same time as something else.

Accretion

Accretion, in geology, is a process by which material is added to a tectonic plate at a subduction zone, frequently on the edge of existing continental landmasses. The added material may be sediment, volcanic arcs, seamounts, oceanic crust, or other igneous features.

In planetary science, accretion refers to the process through which solids clump together to create larger and larger objects, eventually leading to the formation of planets. A disc of gas and minute solid particles with a total mass of about 1% of the gas mass are the starting circumstances. Accumulation must be efficient and quick. Because the original total mass in solids in the solar protoplanetary disk was likely on the order of 300 Earth masses, and the material integrated into the planets is on the order of 100 Earth masses, this method is effective (Rao, 2012). Fast, since the big planets' cores needed to expand to tens of Earth masses in order to capture massive dosages of hydrogen and helium from the disc before it dispersed, which would take millions of years. Depending on the size at which accretion occurs, there is likely more than one accretion process (Figure 4).

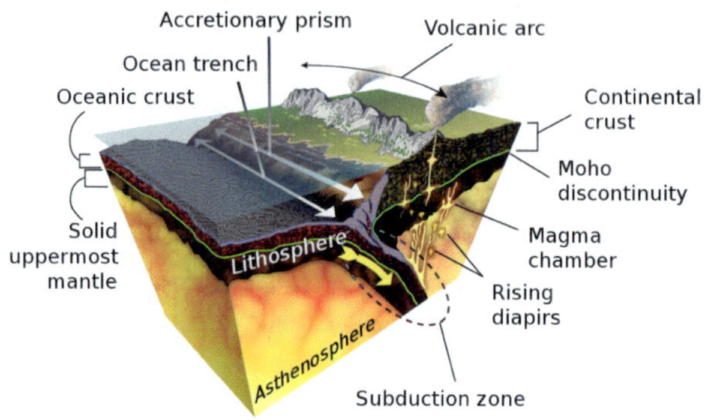

Figure 4. Oceanic-continental convergence and creation of accretionary wedge.

Source: https://en.wikipedia.org/wiki/Accretion_(geology)#/media/File:Subduction-en.svg.

The adhesion of microscopic dust to larger grains and pebbles is one of the initial processes. The production of planetesimals, an intermediate class of objects, is a second process. From planetesimals to planets, a third accretion step is required. Several mechanisms, ranging from collisional coagulation among planetesimals to the accretion of small particles due to gas drag, to large impacts between protoplanets, can be engaged in this step.

Accumulation

Accumulation is the process of water collecting in rivers, lakes, streams, oceans, and other bodies of water. When water condenses and precipitates, it eventually runs off of surfaces and collects again in bodies of water. From there, the water evaporates, and the cycle begins again. Accumulation is important for a number of reasons, both in ecosystems and for humans (Rivet and Kastens, 2012).

Acid

Acids are chemical substances that have a strong taste in water solution, can corrode metals, and can turn certain blue vegetable colors red. Chemical chemicals that are soapy to the touch in solution and convert red vegetable colors blue are known as bases. Acids and bases react with one another to generate salts when combined.

Any substance that tastes sour in water, turns blue litmus paper red, reacts with some metals to liberate hydrogen, combines with bases to form salts, and encourages chemical reactions is classified as an acid (acid catalysis).

Mineral acids include sulfuric acids, nitric acids, hydrochloric acids, and phosphoric acids. Sulfur trioxide, aluminum chloride, and boron trifluoride are examples of nonaqueous acids.

Compounds in the carboxylic acid, sulfonic acid, and phenol groups are examples of organic compounds that are acids.

Acid Rain

Rain or any other form of precipitation that is unusually acidic, meaning it contains high levels of hydrogen ions, is known as acid rain (low pH). The pH of most water, including the water we drink, is neutral, ranging from 6.5 to 8.5, while acid rain has a pH that is lower, ranging from 4–5 on average (Pyle et al., 2018). The lower the pH of acid rain, the more acidic it is. Plants, aquatic animals, and infrastructure can all be harmed by acid rain.

Sulfur dioxide and nitrogen oxide emissions combine with water molecules in the atmosphere to form acids, resulting in acid rain. Since the 1970s, certain governments have worked to minimize the amount of sulfur dioxide and nitrogen oxide released into the environment.

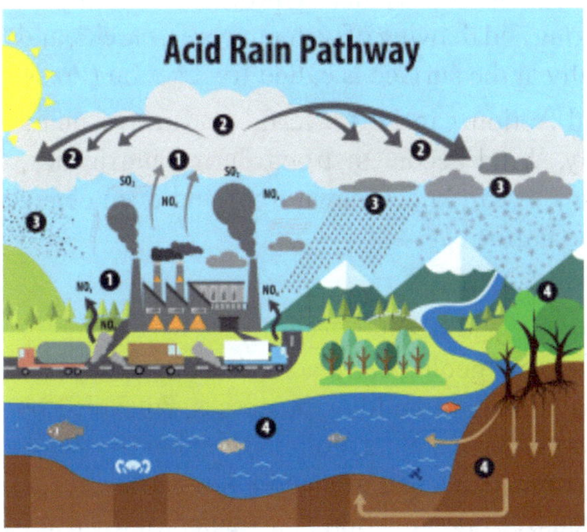

Figure 5. The pathway for acid rain in our environment: (1) Emissions of SO_2 and NOx are released into the air; where (2) the pollutants are transformed into acid particles that may be transported long distances; (3) these acid particles then fall to the Earth as wet and dry deposition (dust, rain, snow, etc.); and (4) may cause harmful effects on soil, forests, streams, and lakes.

Source: https://www.epa.gov/acidrain/what-acid-rain.

Due to widespread research on acid rain beginning in the 1960s and widely broadcast information on its hazardous consequences, these efforts have had beneficial results. Acid rain is mostly caused by anthropogenic sulfur and nitrogen compounds, but nitrogen oxides can also be formed naturally by lightning strikes, and sulfur dioxide is produced by volcanic eruptions (Figure 5).

Acid rain has been found to have negative effects on forests, freshwaters, and soils, killing bacteria, insects, and aquatic life-forms, peeling paint, corroding steel structures such as bridges, and weathering stone buildings and statues, as well as affecting human health.

Acidic

Containing acid.

Action

The occurrence of an event.

- The mechanical weathering process caused by repeated cycles of freezing and thawing of water in pores, cracks, and other openings, usually at the surface is called frost action (Press, 2008).
- Wind Action can be powerful and move objects around quite easily. Wind becomes strong during hurricanes, tornadoes, and storms. It can knock down trees and even change the landscape of the Earth.

Active

When we talk about "geologically active" as it relates to Earth, we know what that means: mountain-building processes, volcanic flows, earthquakes, canyon-creating processes, plate motions, erosion (by wind and water), and so on. These are processes that are studied in geoscience; characterized by action rather than by contemplation or speculation of an active life.

Adiabatic

Adiabatic conditions are those in which there is no total heat transmission across the thermodynamic system's border with the environment. Expansion and compression of gas in a piston-type machine, flow of a fluid medium in heat-insulated pipes, channels, and nozzles, throttling, and setting of turbomachines, and distribution of acoustic and shock waves are all examples of adiabatic processes used in engineering. Adiabatic flow is the movement of a viscous liquid or gas through a heat-insulated tube (Plotnick et al., 2014).

An adiabatic process (Greek: adiábatos, "impassable") is a form of thermodynamic process in which no heat or mass is transferred between the thermodynamic system and its surroundings. An adiabatic process, unlike an isothermal process, only sends energy to the environment as work. The adiabatic process is an important notion in thermodynamics that supports the hypothesis that explains the first rule of thermodynamics.

Because some chemical and physical processes occur too quickly for heat to enter or leave the system, a "adiabatic approximation" can be used.

The adiabatic flame temperature approximation, for example, employs this assumption to calculate the upper limit of flame temperature by assuming combustion does not lose any heat to its surroundings (Figure 6).

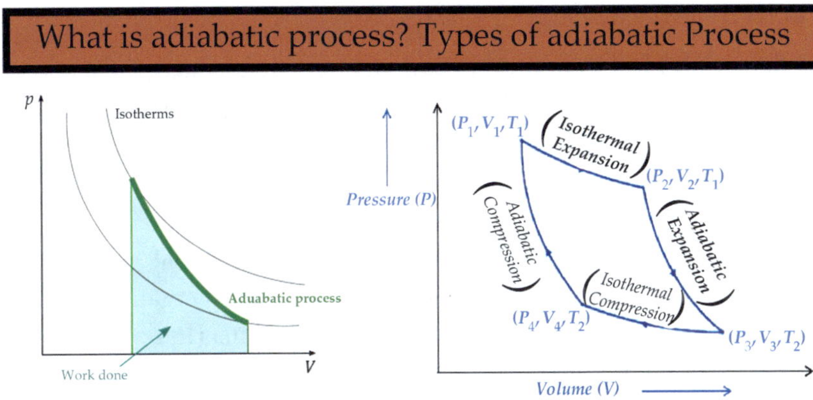

Figure 6. Adiabatic process.

Source: https://engineeringinsider.org/wp-content/uploads/2018/03/341px-Adiabatic.svg_.png.

In meteorology and oceanography, adiabatic cooling causes moisture or salinity to condense, causing the parcel to become oversaturated. As a result, the surplus must be eliminated. The process then becomes a pseudo-adiabatic process, in which the condensing liquid water or salt is thought to be eliminated by idealized instantaneous precipitation upon creation. Because a compressed parcel becomes warmer and remains undersaturated, the pseudoadiabatic process is only described for expansion (Pennington et al., 2020).

Adjustment

The act of adjusting or the state of being adjusted. Earth is continually moving and changing, even if slowly. Over millions of years, temperatures rise and fall in cycles. The last ice age happened only 16,000 years ago, when two-mile-thick sheets of ice blanketed much of the Northern Hemisphere. Though the ice has long since melted, the Earth beneath and around it continues to rise and fall in response to its ice-age burden (Perdigão et al., 2020).

Glacial isostatic adjustment refers to the continuing movement of land. The following is how it works: Consider lying down on a plush mattress and then rising from the same position. You can see a puffed-up region around the indentation where the mattress raised, as well as an indentation in the mattress where your body had been. When you get out of bed, it takes a little while for the mattress to relax and return to its normal shape.

Advection

Advection is the horizontal or lateral movement of mass, heat, or other properties. As a result, winds that blow across the Earth's surface are advection air movements. Advection occurs in the ocean as well, in the form of currents. Geologists are currently debating the existence and role of significant advective processes in the Earth's mantle.

The bulk movement of air is driven by differential pressures and temperatures as it seeks equilibrium (the lowest energy state). Winds that are more aggressive go from hotter to cooler locations (Parkes and Morris, 2001). Convection, on the other hand, is the vertical movement of material or heat transfer that emerges as air currents. As a result, winds are caused by advection, whereas air currents are caused by convection.

The atmosphere, despite being in a gaseous form, exhibits fluid-like movements. Because advection is frequently more prominent in the movement of fluids, this is an important issue when discussing advection. For example, advection also takes place in the oceans where advection is broadened to include the lateral (horizontal) transfer of not only fluid mass and heat, but of other properties such as oxygen content and salinity.

Despite the fact that advection processes are significant heat equilibration mechanisms in both the atmosphere and the oceans, the speed and volume of material carried differ greatly. Heat flux (the pace of heat transport) determines the magnitude of heat transference, and flux, in turn, links the movement of heat energy in terms of area and time (Özyurt et al., 2021).

Because wind currents are significantly faster (at a greater rate) than ocean currents, but ocean currents move much heavier amounts of molecules, both processes contribute roughly equally (Figure 7).

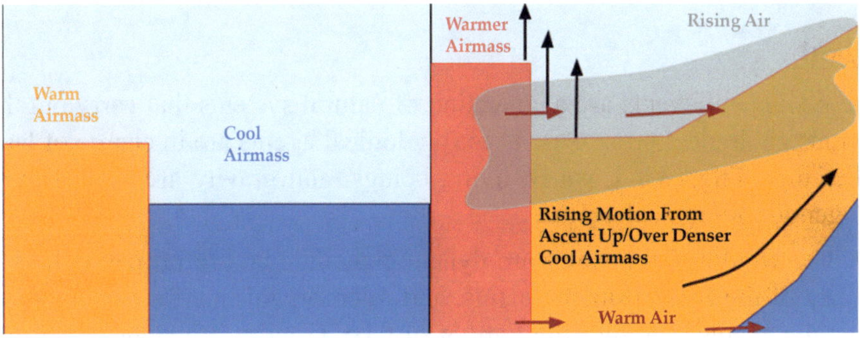

Figure 7. Advection.

Source: https://blog.weather.us/advection-what-is-it-and-why-is-it-important/.

The production of advection fog is likewise caused by advection. Advection fog develops when the atmosphere is relatively stable, preventing moist (humid) air near the surface from mixing vertically with a layer of dry air underlying it. As warm and wet air flows horizontally along the cooler surface, the air near the surface cools to its dew point, forming advection fog.

Aerosol

A suspension of small solid particles or liquid droplets in air or another gas is known as an aerosol. Aerosols can be either natural or man-made. Fog or mist, dust, woodland exudates, and geyser steam are examples of natural aerosols. Particulate air pollution and smoke are examples of anthropogenic aerosols (Orion, 2019). The liquid or solid particles are typically smaller than 1 m in diameter; larger particles with a high settling speed make the combination a suspension, but the distinction is not always evident. In general, an aerosol spray that distributes a consumer product from a can or similar container is referred to as an aerosol. Pesticide dispersal, medical treatment of respiratory disorders, and combustion technology are some of the other technological applications of aerosols. An anthropogenic aerosol is sucked into the lungs when a person inhales the contents of a vape pen or e-cigarette. Diseases can also spread by means of small droplets in the breath, also called aerosols (or sometimes bioaerosols).

Aerosol science covers the generation and removal of aerosols, technological application of aerosols, effects of aerosols on the environment and people, and other topics.

Agent

The geological agents are a collection of natural systems that carry out the various geological processes. These geological agents are in charge of land relief modeling. Wind, water, living beings, and gravity are examples of exogenous geological agents.

The wind and the water are dynamic agents that are propelled by the energy of the Sun. Animals, plants, and microorganisms, for example, are dynamic agents that alter the Earth's surface. Gravity is a static agent that is always present in all of the dynamic agents' processes (Oreskes, 2015).

Wind, water, and ice erode and shape the land. Volcanic activity and earthquakes alter the landscape in a dramatic and often violent manner. And on a much longer timescale, the movement of Earth's plates slowly reconfigures oceans and continents.

Aggregate

Aggregate can have three different meanings in the Earth sciences. An aggregation is a mass of mineral crystals, mineraloid particles, or rock particles in mineralogy and petrology. Dolomite, which is an aggregate of dolomite crystals, and rock gypsum, which is an aggregate of gypsum crystals, are two examples. Lapis lazuli is a form of rock made up of crystals of a variety of minerals such as lazurite, pyrite, phlogopite, calcite, potassium feldspar, wollastonite, and a few sodalite group minerals (Oreskes, 2013).

An aggregate (also known as construction aggregate) is sand, gravel, or crushed rock that has been mined or quarried for use as a building material in the construction industry.

An aggregate is a mass of soil particles in pedology. If the aggregate formed naturally, it is known as a ped; if it was created artificially, it is known as a clod.

Air Mass

A huge volume of air in the atmosphere that is relatively uniform in temperature and moisture is referred to as an air mass. Air masses can stretch thousands of kilometers across the Earth's surface and reach up to

16 kilometers (10 miles) into the stratosphere. Source regions are where air masses from across wide surfaces with constant temperatures and humidity. Low wind speeds allow air to remain fixed over lengthy periods of time, allowing it to absorb the characteristics of the source region, such as heat or cold (Odenwald, 2020). Winds transport meteorological conditions (hot or cold, dry, or moist) from one location to another when they move air masses. When the air mass arrives in a new area, it may collide with another air mass with a different temperature and humidity. This has the potential to cause a severe storm (Figure 8).

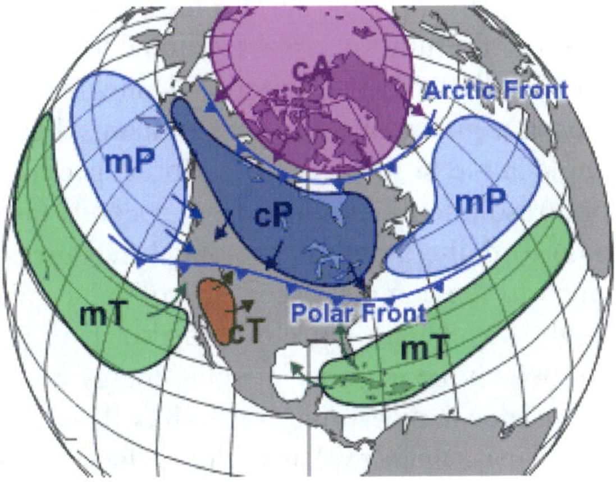

Figure 8. Different air masses which affect North America as well as other continents, tend to be separated by frontal boundaries.

Source: https://upload.wikimedia.org/wikipedia/commons/7/76/Airmassesorigin.png.

Air masses are classified by meteorologists based on where they form over the Earth. Air masses are divided into four categories: arctic, tropical, polar, and equatorial. Arctic air masses are extremely cold and form in the Arctic area. Tropical air masses are warm air masses that originate in low-latitude locations. Polar air masses are chilly and form in high-latitude locations. The warm equatorial air masses form around the Equator.

The formation of air masses over land or sea is also used to identify them. Maritime air masses are humid and originate over water. Continental air masses are dry air masses that form over land (Neeck and Volz, 2011).

As a result, an air mass forming over northern Canada is known as a continental polar air mass, and it is cold and dry. A maritime tropical air mass is a warm and humid air mass that forms over the Indian Ocean.

Albedo

Albedo is a measure of diffuse solar radiation reflection out of total solar radiation, and it is measured on a scale of 0 to 1, with 0 equating to a black body that absorbs all incident radiation and 1 corresponding to a body that reflects all incident radiation.

The ratio of radiosity Je to irradiance Ee (flux per unit area) received by a surface is known as surface albedo. The fraction reflected is governed by the spectral and angular distribution of solar radiation reaching the Earth's surface, as well as the qualities of the surface itself. These variables change with the composition of the atmosphere, geographic location, and time (see position of the Sun). Albedo is the directional integration of reflectance overall solar angles in a given time, whereas bi-hemispheric reflectance is calculated at a single angle of incidence (i.e., for a specific position of the Sun). The temporal resolution may range from seconds (as obtained from flux measurements) to daily, monthly, or annual averages (Movva et al., 2012). Albedo refers to the complete spectrum of solar radiation unless it is stated for a single wavelength (spectral albedo). It is frequently given for the spectrum in which the majority of solar energy reaches the surface (between 0.3 and 3 m) due to measuring constraints. This spectrum includes visible light (0.4–0.7 m), which explains why surfaces with a low albedo (e.g., trees) appear dark and surfaces with a high albedo appear brilliant (e.g., snow reflects most radiation).

Albedo is a key concept in climatology, astronomy, and environmental management (for example, in the leadership in energy and environmental design (LEED) program for building sustainability rating). Because of cloud cover, the Earth's average albedo from the upper atmosphere, or planetary albedo, is 30–35%, but it varies greatly regionally over the surface due to diverse geological and environmental factors.

Alloy

A combination or metallic-solid solution made up of two or more elements is known as an alloy. Materials such as brass, pewter, phosphor bronze, amalgam, and steel are examples of alloys. The microstructure of complete solid solution alloys is one solid phase. Depending on the thermal history, partial solutions produce two or more phases that may or may not be homogenous in distribution. The properties of an alloy are frequently distinct from those of its constituent elements (Müller et al., 2019).

The mass of alloy ingredients is commonly measured. Depending on its atomic configuration, an alloy is usually classed as either substitutional or interstitial. Each element's atoms can occupy the same positions as their counterparts in a substitutional alloy. The atoms in interstitial alloys do not occupy the same locations. Alloys can be homogeneous (consisting of a single-phase), heterogeneous (containing two or more phases), or intermetallic (consisting of two or more phases) (where there is no distinct boundary between phases).

Alluvial

A deposit of sand, mud, etc., formed by flowing water. The sedimentary matter deposited thus within recent times, especially in the valleys of large rivers. River-deposited material is known as alluvial deposit. It is made up of silt, sand, clay, and gravel, as well as a significant amount of organic materials. Alluvial deposits are most common in the lower reaches of a river's course, generating floodplains and deltas, although they can arise anywhere a river exceeds its banks or has its flow restricted (Metz, 2008). The deltas of the Mississippi, Nile, Ganges, and Brahmaputra, as well as the Huang (Yellow) rivers, produce extremely fertile soils. They contain the majority of the world's tin ore, as well as gold, platinum, and gemstones in some areas.

Alluvial Fan

A shallow cone-shaped accumulation of sediments with its apex at a point source of sediments, such as a narrow canyon emerging from an escarpment, is known as an alluvial fan.

Figure 9. An alluvial fan spreads out in Death Valley.

Source: https://www.nationalgeographic.org/encyclopedia/alluvial-fan/.

They are common in hilly terrain in dry to semiarid climates, although they can also be found in more humid conditions with heavy rainfall and in locations where contemporary glaciation has occurred. They cover an area of less than 1 square kilometer (0.39 square mile) to about 20,000 square kilometers (7,700 sq mi). When flow emerges from a constricted channel and is free to spread out and permeate the surface, alluvial fans form. This decreases the flow's carrying capacity, resulting in sediment deposition. Infrequent debris flows or one or more ephemeral or perennial streams might cause the flow (Figure 9).

Alluvial fans can be found throughout the geologic record, such as in the eastern North American Triassic basins and the New Red Sandstone of South Devon. The biggest accumulations of gravel in the geologic record are expected to be found in such fan deposits (Li and Li, 2016).

On the Indo-Gangetic plain, along the Himalayan Mountain range, some of the largest alluvial fans may be found. As happened on the Kosi River fan in 2008, a shift in the feeder channel (a nodal avulsion) might result in catastrophic flooding.

Alluvium

Alluvium (from the Latin alluvius, from allure, "to wash against") is loose clay, silt, sand, or gravel deposited by running water in a stream bed, floodplain, alluvial fan or beach, or other similar locations. Alluvium can also be referred to as an alluvial deposit. Alluvium is a geologically young material that has not yet been cemented into solid rock. Alluvium does not refer to sediments deposited underwater in seas, estuaries, lakes, or ponds.

Alluvium from floodplains can be extremely productive, and it has supported some of humanity's first civilizations.

Alpha

Alpha is the first letter of the Greek alphabet. In the system of Greek numerals, it has a value of one. Alpha particles, also known as alpha rays or alpha radiation, are made up of two protons and two neutrons bonded together to form a helium-4 nucleus-like particle. They are usually made during the alpha decay process, although they can also be made in other ways. Alpha particles are named from the Greek alphabet's initial letter. The alpha particle is represented by the symbol or 2+. They are also written as He^{2+} or 4 $2He^{2+}$, denoting a helium ion with a +2 charge, because they are similar to helium nuclei (missing its two electrons). The alpha particle becomes a regular (electrically neutral) helium atom 4 2He after gaining electrons from its environment (Ledley et al., 2011).

The alpha particle becomes a regular (electrically neutral) helium atom 4 2He after gaining electrons from its environment.

Alpha decay, type of radioactive disintegration in which some unstable atomic nuclei dissipate excess energy by spontaneously ejecting an alpha particle. Because alpha particles have two positive charges and a mass of four units, their emission from nuclei produces daughter nuclei having a positive nuclear charge or atomic number two units less than their parents and a mass of four units less.

Alpha Centauri

At 4.37 light-years (1.34 parsecs) from the Sun, Alpha Centauri is a gravitationally locked system of the closest stars and exoplanets to our Solar System. The name is shortened Alpha Cen or Cen and is Latinized from Centauri. Centauri A (officially Rigil Kentaurus), Centauri B (officially Toliman), and Centauri C (officially Rigil Kentaurus) make up the triple star system (officially Proxima Centauri). Sun-like stars Alpha Centauri A and B (Class G and K, respectively) combine to form the binary star Alpha Centauri AB. To the naked eye, the two major components appear to be a single star with an apparent magnitude of 0.27, the brightest star in Centaurus and the third brightest in the night sky, behind Sirius and Canopus (Lewis, 2008).

Alpha Centauri A is larger and brighter than the Sun, with 1.1 times its mass and 1.519 times its brightness, while Alpha Centauri B is smaller and colder, with 0.907 times the Sun's mass and 0.445 times its luminosity. With an orbital period of 79.91 years, the duo orbits a common center. The distance between A and B fluctuates from 35.6 AU (astronomical units), or about the distance between Pluto and the Sun, to 11.2 AU, or about the distance between Saturn and the Sun, due to their eccentric elliptical orbit (Figure 10).

Alpha Centauri C, often known as Proxima Centauri, is a dim red dwarf (Class M). Proxima Centauri is the closest star to the Sun at a distance of 4.24 ly (1.30 pc), somewhat closer than Alpha Centauri AB, albeit it is not visible to the human eye. Proxima Centauri and Alpha Centauri AB are currently separated by around 13,000 AU (0.21 ly), equivalent to about 430 times the radius of Neptune's orbit (Košler, 2007).

Figure 10. Alpha, Beta, and Proxima Centauri.

Source: https://en.wikipedia.org/wiki/File:Alpha,_Beta_and_Proxima_Centauri_(1).jpg.

Alternate

To interchange repeatedly and regularly with one another in time or place; rotate (usually followed by with): Day alternates with night. To change back and forth between conditions, states, actions, etc.

Alternative Energy

Alternative energy refers to non-fossil fuel energy sources. This encompasses all renewable energy sources as well as nuclear power. Nuclear power is not a renewable energy source. Renewable energy is generated from non-depleting or replenishable resources that can be replenished during a human lifetime. Nuclear power is made from nonrenewable elements such as uranium and thorium, which are mined (Kleinhans et al., 2010).

Alternative energy sources are generally seen favorably because they do not produce greenhouse gases directly, but all forms of energy have their own costs, advantages, and tradeoffs.

- Hydrogen gas;
- Tidal energy;
- Biomass energy;
- Wind energy;
- Geothermal power;

- Natural gas;
- Biofuels.

Altitude

Altitude or height is a distance measurement, usually in the vertical or "up" direction, between a reference datum and a point or object. The exact definition and reference datum varies according to the context.

Altostratus

Altostratus is a middle-altitude cloud genus belonging to the stratiform physical category characterized by a generally uniform gray to bluish-green sheet or layer. It is lighter in color than nimbostratus and darker than high cirrostratus (Klippel et al., 2019.

Amplify

- To increase in number, effect, size, volume, amplitude of, etc.;
- To make larger, greater, stronger; to enlarge.

Anaerobic

The term anaerobic refers to biological processes that take place in the absence of molecular oxygen (typically the release of energy from nutrients).

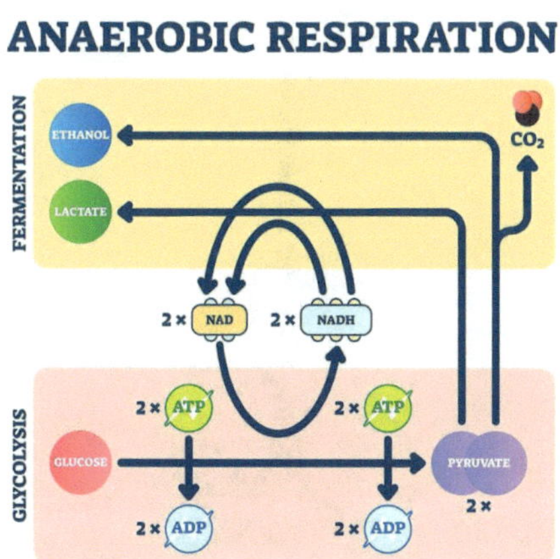

Figure 11. Anaerobic respiration.

Source: https://biologydictionary.net/anaerobic-respiration/.

Prokaryotic bacteria (organisms whose genetic material is not housed within a specific nuclear membrane) most likely evolved in environments with limited or no oxygen (Hundman and Mattmann, 2017). The organisms derived energy from organic substances that were devoid of oxygen (that is, by anaerobic respiration). Many species alive today get their energy from nutrients in an aerobic (oxygen-rich) manner, though only a few respire. Aerobic respiration produces a large amount of energy from nutrients, whereas anaerobic respiration produces a little amount (Figure 11).

Analysis

To get a better grasp of a complex topic, material or substance, analysis is the process of breaking it down into smaller bits. Although analysis as a formal idea is a new creation, it has been used in the study of mathematics and logic since before Aristotle.

Anemometer

An anemometer is a device that calculates wind speed and pressure. Anemometers are significant implements for meteorologists, who examine weather trends. The anemometer estimates wind speed and communicates wind speed details to the regulator.

Figure 12. A hemispherical cup anemometer of the type invented in 1846 by John Thomas Romney Robinson.

Source: https://en.wikipedia.org/wiki/Anemometer.

Aneroid

Using no liquid specifically: operating by the effect of outside air pressure on a diaphragm forming one wall of an evacuated container aneroid barometer (Guo et al., 2016). An aneroid barometer is a metal box, containing copper and beryllium, that measures atmospheric pressure.

Angular

The angle between two sightlines, or between two-point objects, as seen from an observer is known as angular separation, apparent distance, or apparent separation.

Angular distance is found in all-natural sciences and mathematics (particularly geometry and trigonometry) (e.g., astronomy and geophysics). It appears alongside angular velocity, angular acceleration, angular momentum, moment of inertia, and torque in classical mechanics of rotating objects (Bohle, 2015).

Anion

An anion is an ion with more electrons than protons, giving it a net negative charge (since electrons are negatively charged and protons are positively charged).

Antecedent

Preceding in time or order; previous or pre-existing. An antecedent stream is a stream that maintains its original course and pattern despite the changes in underlying rock topography. A stream with a dendritic drainage pattern, for example, can be subject to slow tectonic uplift (Goldstein et al., 2017). However, as the uplift occurs, the stream erodes through the rising ridge to form a steep-walled gorge. The stream thus keeps its dendritic pattern even though it flows over a landscape that will normally produce a trellis drainage pattern.

Anticlinal

An anticline is a form of fold in structural geology that has an arch-like structure with the oldest beds near its center, whereas a syncline is the inverse of an anticline. The hinge or crest is the area where the curvature is greatest in a typical anticline, and the limbs are the sides of the fold that descend away from the hinge. Anticlines are distinguished from antiforms by a series of rock layers that become older as they travel toward the fold's center. As a result, the term antiform should be used when age correlations between diverse rock strata are uncertain.

Anticyclone

An anticyclone is a weather phenomenon defined as a large-scale circulation of winds around a central region of high atmospheric pressure, clockwise in the Northern Hemisphere and counterclockwise in the Southern Hemisphere as viewed from above (Butler, 2016).

Anticyclonic

Strong yet vertically shallow high-pressure systems result from the passage of continental arctic air masses to lower latitudes. Sharp temperature inversions at the surface can result in patches of prolonged stratocumulus or stratus cloud, often known as anticyclonic gloom. The sort of weather caused by an anticyclone is determined by its origin. Because they take up moisture as they move across the warmer oceans, extensions of the Azores high pressure, for example, may generate anticyclonic darkness in the winter. Because high pressures build to the north and flow south, they generally provide clear weather because they are chilled (rather than warmed) at the base, preventing clouds from forming.

Anvil

Anvil presses are used in materials science and geology to synthesize and study various phases of materials under extreme pressure, as well as for the industrial production of valuable minerals, particularly synthetic diamonds, because they mimic the pressures and temperatures found deep within the Earth.

In geology, engineering, and materials science experiments, a diamond anvil cell (DAC) is a high-pressure instrument. It allows a small (sub-millimeter-sized) piece of material to be compressed to extremely high pressures, typically approximately 100–200 gigapascals, while pressures of up to 770 gigapascals have been achieved (7,700,000 bars or 7.7 million atmospheres) (Escorihuela, 2018).

Aphelion

Aphelion is the farthest point in the orbit of a planet, comet, or other body from the Sun in astronomy. The Earth is about 4,800,000 km (3,000,000 miles) further from the Sun during its aphelion in early July than it is during its perihelion in early January (Duffy, 2014). The phrases apogee (for farthest point from Earth) and a posaturnium are used to describe the most distant orbital point of entities traveling around centers of force other than the Sun (for farthest point from Saturn). The Moon's apogee, for example, is around 50,000 kilometers (31,000 miles) farther from Earth than its closest orbital point (perigee). The term apocenter refers to the point in a body's orbit that is the furthest away from the body it circles (Figure 13).

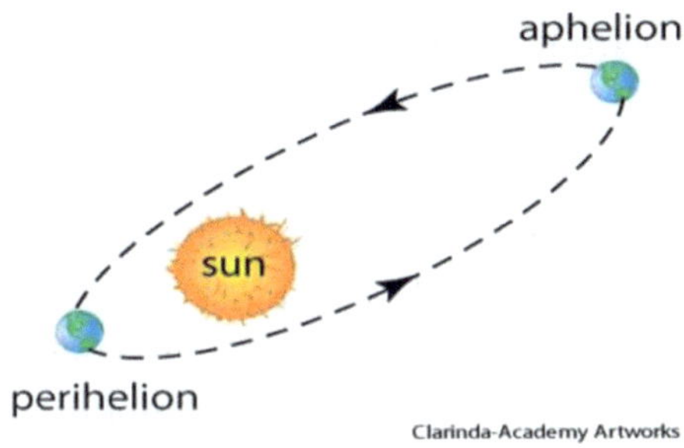

Figure 13. Aphelion.

Source: https://qsstudy.com/geology/aphelion.

Apogee

The point in the orbit of a heavenly body, especially the moon, or of a man-made satellite at which it is farthest from the Earth. The highest or most distant point; climax.

Apparent

Clearly visible or understood; obvious. An optical illusion in which stationary objects viewed in quick succession or in relation to moving objects appear to be in motion is called apparent movement (Clowes, 2010). Apparent magnitude is the luminosity of a celestial body (such as a star) as observed from the Earth.

Aquifer

A subterranean layer of water-bearing porous rock, rock fractures, or unconsolidated materials is known as an aquifer (gravel, sand, or silt). A water well can be used to harvest groundwater from aquifers. Hydrogeology is the study of water movement in aquifers and the characterization of aquifers. Aquitard, a low-permeability bed along an aquifer, and aquiclude (or aquifuge), a solid, impermeable area below or covering an aquifer, the pressure of which could form a restricted aquifer, are concepts that are related (Clary and Wandersee, 2011).

Saturated versus unsaturated aquifers; aquifers against aquitards; confined versus unconfined aquifers; isotropic versus anisotropic aquifers; porous, karst, or fractured aquifers; transboundary aquifers Subsidence, saltwater intrusion, salination, and groundwater pollution are all challenges for using groundwater from aquifers.

Argon

Argon is a chemical element with atomic number 18 and the symbol Ar. It is a noble gas and belongs to group 18 of the periodic table. At 0.934%, argon is the third most abundant gas in the Earth's atmosphere (9,340 ppmv). It is more than twice as plentiful as water vapor (approximately 4,000 ppmv on average, but fluctuates widely), 23 times as abundant as carbon dioxide (CO_2) (400 ppmv), and more than 500 times as abundant as neon (18 ppmv). Argon is the most abundant noble gas in the Earth's crust, accounting for 0.00015% (Figure 14) (Cossu et al., 2010).

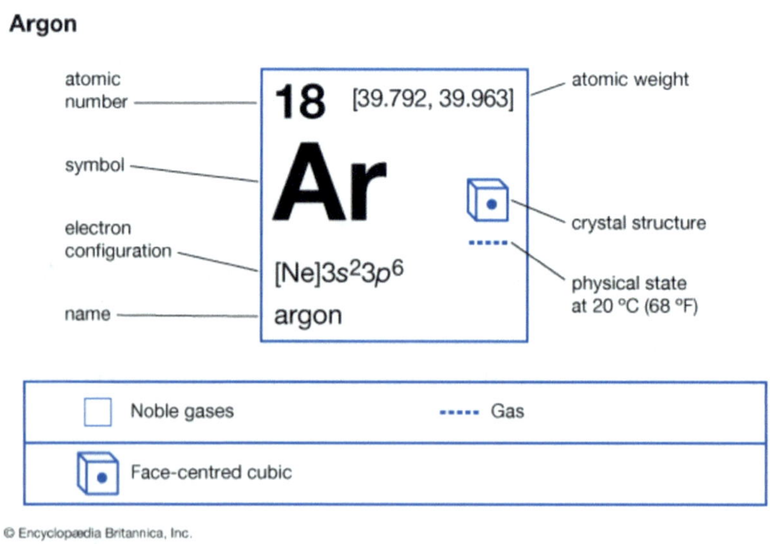

Figure 14. Argon.

Source: https://www.britannica.com/summary/argon-chemical-element.

Radiogenic argon-40, generated from the breakdown of potassium-40 in the Earth's crust, accounts for nearly all of the argon in the atmosphere. Because it is the most easily created by stellar nucleosynthesis in supernovas, argon-36 is by far the most prevalent argon isotope in the universe.

Artesian

The word artesian, properly used, refers to situations where the water is confined under pressure below layers of relatively impermeable rock.

Artesian – (of water) rising to the surface under internal hydrostatic pressure; Artesian well, a well from which water flows under natural pressure without pumping. It is dug or drilled wherever a gently dipping, permeable rock layer (such as sandstone) receives water along its outcrop at a level higher than the level of the surface of the ground at the well site (Cerase, 2019).

Asteroid

Asteroids are small planets in our Solar System. Historically, these phrases have been used to describe any astronomical object orbiting the Sun that did not resolve into a disc in a telescope and did not exhibit active comet-like characteristics such as a tail. As minor planets in the outer Solar System with volatile-rich surfaces comparable to comets were discovered, they were distinguished from the objects found in the main asteroid belt. As a result, the term "asteroid" now applies to all minor planets in the inner Solar System, including those in Jupiter's orbit. Planetoids are the names given to larger asteroids.

Asthenosphere

The asthenosphere is the Earth's upper mantle's highly viscous, mechanically weak, and ductile area. It is located in the lithosphere, between 80 and 200 kilometers (50 and 120 miles) below the surface. The LAB is the name given to the lithosphere-asthenosphere border. Although parts of the asthenosphere's regions are molten, the asthenosphere is almost solid (e.g., below mid-ocean ridges (MORBs)). The asthenosphere's lower limit within the mantle is not precisely defined. The asthenosphere's thickness is mostly determined by temperature (Britner, 2008). The rheology of the asthenosphere, on the other hand, is affected by the rate of deformation, suggesting that the asthenosphere might form as a result of a high rate of deformation. The asthenosphere could reach a depth of 700 kilometers in some areas (430 mi). It is thought to be the source of basalt from the MORB (Figure 15).

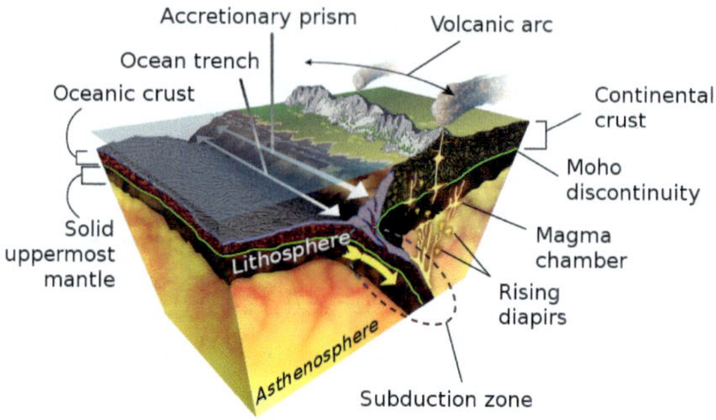

Figure 15. The asthenosphere showed at a subduction boundary.

Source: https://en.wikipedia.org/wiki/File:Subduction-en.svg.

Astronomer

An astronomer is a scientist in the field of astronomy who focuses their studies on a specific question or field outside the scope of Earth. They observe astronomical objects such as stars, planets, moons, comets, and galaxies – in either observational (by analyzing the data) or theoretical astronomy (Zhuang et al., 2019).

Examples of topics or fields astronomers study include planetary science, solar astronomy, the origin or evolution of stars, or the formation of galaxies. Related but distinct subjects are like physical cosmology, which studies the Universe as a whole.

Astronomy

Astronomy is a branch of natural science concerned with the study of celestial objects and events. It explains their origin and evolution using mathematics, physics, and chemistry. Planets, moons, stars, nebulae, galaxies, and comets are among the objects of interest. Supernova explosions, gamma-ray bursts, quasars, blazars, pulsars, and cosmic microwave background radiation are all relevant phenomena. Astronomy, in general, is the study of anything that originates outside of Earth's atmosphere. Cosmology is an astronomical branch that investigates the entire cosmos (Yu and Gong, 2012).

One of the oldest scientific sciences is astronomy. The night sky was meticulously observed by early civilizations in recorded history.

Babylonians, Greeks, Indians, Egyptians, Chinese, Maya, and many more ancient indigenous peoples of the Americas are among them.

Atmosphere

An atmosphere is a layer (or layers) of gas that surrounds a planet and is held in place by the planetary body's gravity. When the gravity is high and the temperature of the atmosphere is low, a planet preserves its atmosphere. The outer part of a star, which includes the layers above the opaque photosphere, is known as the stellar atmosphere; low-temperature stars may have outer atmospheres containing complex molecules.

The atmosphere of Earth is made up of 78% nitrogen, 21% oxygen, 0.9% argon, 0.04% CO_2, and trace gases. The majority of animals breathe oxygen; lightning and bacteria fix nitrogen to produce ammonia, which is required to synthesize nucleotides and amino acids; and plants, algae, and cyanobacteria use CO_2 for photosynthesis (Witze, 2013). The layered composition of the atmosphere minimizes the harmful effects of sunlight, ultraviolet (UV) radiation, the solar wind, and cosmic rays to protect organisms from genetic damage. The current composition of the atmosphere of the Earth is the product of billions of years of biochemical modification of the paleoatmosphere by living organisms.

Atmospheric Pressure

The air surrounding you is heavy, and it pushes against everything it comes into contact with. Atmospheric pressure, or air pressure, is the name given to this pressure. The air above a surface exerts a force on it as gravity draws it to Earth.

A barometer is often used to measure atmospheric pressure. As the weight of the atmosphere changes, a column of mercury in a glass tube rises or falls in a barometer. The height to which the mercury rises is used by meteorologists to describe atmospheric pressure (Figure 16).

At 15°C, an atmosphere (atm) is a unit of measurement equal to the average air pressure at sea level (59°F). One atmosphere contains 1,013 millibars of mercury, or 760 millimeters (29.92 inches) (Wilkinson et al., 2020). As height rises, atmospheric pressure decreases. Denali, Alaska, has about half the atmospheric pressure of Honolulu, Hawaii. Honolulu is a city at sea level. Denali, also known as Mount McKinley, is the highest peak in North America.

Figure 16. The atmospheric gases around Earth scatter blue light (shorter wavelengths) more than light toward the red end (longer wavelengths) of the visible spectrum; thus, a blue glow over the horizon is seen when observing Earth from outer space.

Source: https://en.wikipedia.org/wiki/Atmosphere#/media/File:Top_of_Atmosphere.jpg.

As the pressure decreases, the amount of oxygen available to breathe also decreases. At very high altitudes, atmospheric pressure and available oxygen get so low that people can become sick and even die.

Atoll

An atoll, also known as a coral atoll, is a ring-shaped coral reef with a coral rim that partially or completely encircles a lagoon. On the rim, there could be coral islands or cays. Atolls are coral-rich islands found in warm tropical or subtropical oceans and seas. The Pacific Ocean is home to the majority of the world's 440 atolls (Wallace-Wells, 2019).

The subsidence and antecedent karst hypotheses, both widely utilized, have been used to explain the formation of atolls. The construction of an atoll is explained by the subsidence of a volcanic island around which a coral fringing reef has formed, according to Charles Darwin's subsidence model. The volcanic island becomes extinct and degraded throughout geologic time as it sinks fully beneath the ocean's surface. The coral fringing reef becomes a barrier reef that is detached from the island as the volcanic island falls.

Atom

The atom is the smallest unit of matter that may be divided without releasing electrically charged particles. It is also the smallest unit of matter with chemical element features. As a result, the atom is the fundamental unit of chemistry.

The majority of the atom is made up of empty space. The rest is made up of a cloud of negatively charged electrons surrounding a positively charged nucleus of protons and neutrons. In comparison to electrons, which are the lightest charged particles in nature, the nucleus is small and dense. Electric forces attract electrons to any positive charge; in an atom, electric forces bind electrons to the nucleus (Toth et al., 2014).

The electrons of an atom behave like particles orbiting the nucleus in various ways. In others, the electrons act like waves around the nucleus that are frozen in place. Orbitals are wave patterns that describe the distribution of individual electrons. These orbital features have a significant impact on an atom's behavior, and orbital groups known as shells dictate an atom's chemical properties.

Aurora

A natural light shows that shimmers in the sky is known as an aurora. Like softly blowing curtains, colorful blue, red, yellow, green, and orange lights fluctuate and change shape. Auroras may only be seen at night and are frequently found in the lower polar areas.

Near the Arctic and Antarctic Circles, which are about 66.5° north and south of the Equator, auroras can be seen virtually every night (Tollefson, 2015). The spectacle is known as aurora borealis, or northern lights, in the north. It is known as the aurora australis, or southern lights, in the southern hemisphere.

The Sun is the source of the activity that causes auroras. The solar wind is a continuous stream of ions emanating from the Sun's surface (Figure 17).

Although the magnetosphere blocks most of the solar wind, some of the ions are temporarily held in ring-shaped holding zones around the globe. These zones are located around the Earth's geomagnetic poles in the ionosphere, a portion of the atmosphere. The skewed axis of the Earth's magnetic field is shown by the geomagnetic poles. They are around 1,300 kilometers (800 miles) away from the geographic poles, but they are traveling slowly (Thessen, 2016).

Figure 17. Solar wind collides with the Earth's magnetic field when it approaches the planet. The solar wind would blast away Earth's fragile atmosphere without this magnetic field protecting it, putting all life on the planet in jeopardy. The magnetosphere blocks most of the solar wind, but the ions that are driven around the planet continue to move farther into the solar system.

Source: https://en.wikipedia.org/wiki/Aurora#/media/File:Aurora_Borealis_and_Australis_Poster.jpg.

Avalanche

A mass of snow, rock, ice, soil, and other debris flows down a mountainside as an avalanche. Landslides are a type of avalanche that occurs when rocks or soil avalanche. Avalanches, especially snowslides, can sweep downhill quicker than the fastest skier.

When an unstable mass of snow breaks away from a slope, a snow avalanche occurs. As it goes downhill, the snow increases up pace, forming a river of snow and a cloud of frozen particles that rises high into the air. As it rushes downward, the moving mass scoops up even more snow. An avalanche that has fully built can weigh up to a million tons. It has a top speed of 320 kilometers per hour (200 miles per hour).

Layers in a snowpack slip off, causing avalanches. A snowpack is simply the accumulation of layers of snow in a certain place, such as the slope of a mountain (Telling et al., 2017).

During the winter, a dozen-of-meter-thick snowpack is formed by repeated snowfalls. The thickness and texture of the layers vary.

It is possible that the bonds between the layers of a snowpack are weak. Melted snow that refreezes can result in a smooth coating of ice forming on the layer's surface. A new snowfall may not adhere to this slick coating, causing it to slide away. Melted snow can seep through a snowpack during the spring thaw, making the below layer's surface slick. The top layers of a snowfall can readily be sent hurtling downhill by added weight or vibration.

Axis

An axis is an invisible line around which a rotating or spinning object revolves. A minuscule particle, smaller than a single atom, might be the object. It might also be a star the size of 1,000 suns.

An object's axis, in either scenario, passes via its center of mass, or barycenter. The center of mass of an object is the position at which an external force operating on the object operates as if the thing were located exactly at that point—the point at which the object seems "balanced." The Earth's center of mass changes over time. Ocean tides change the planet's center of mass, but not enough to cause the planet's axis to shift dramatically (Showstack, 2008).

Our solar system's planets spin on their axes. As a result, each planet has a North and South Pole, which are the sites where an axis intersects the surface. The time it takes for a planet or other celestial object to complete one spin around its axis is called its rotation period. Earth's rotation period is about 24 hours or one day.

B

Barometer

The term "barometer" refers to a device that measures atmospheric pressure. A barometer can also be used to estimate height because air pressure changes with distance above or below sea level. Mercury and aneroid barometers are the two most common variants (Figure 18).

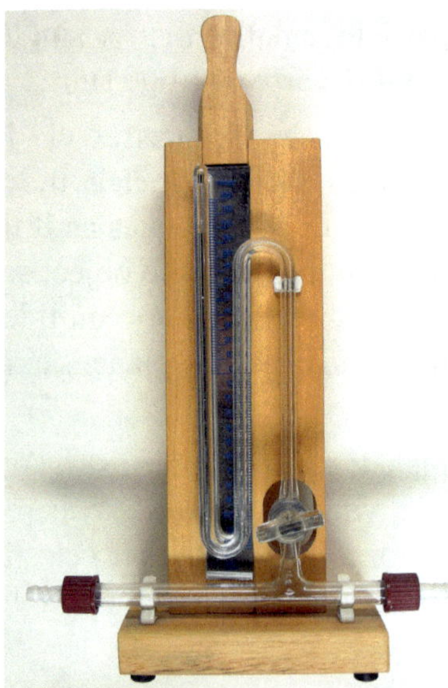

Figure 18. Barometer.

Source: https://www.britannica.com/technology/barometer.

The mercury barometer works by balancing ambient pressure on a mercury column whose height can be precisely determined. Mercury barometers are frequently calibrated for ambient temperature and the local gravity value to improve their accuracy. Pounds per square inch, dynes per square centimeter, newtons per square meter (the SI unit called the pascal), inches, centimeters, or millimeters of mercury, and millibars are all typical pressure units (1 millibar equals 1,000 dynes per square centimeter, 100 pascals, or 0.75 millimeters of mercury). At sea level, atmospheric pressure is around 14.7 pounds per square inch, or 30 inches (760 mm) of mercury, 1,013.2 millibars, or 101,320 pascals (Schmitz, 2011).

Barrier Island

A barrier island is a coastal landform made up of an elongated thin strip of unconsolidated sediment (mostly sand) that runs parallel to the mainland shore and is separated from it by a lagoon, salt marsh, or bay.

Barrier islands can be found all over the planet. However, there are differing viewpoints on what constitutes a barrier island. Barrier island definitions that are descriptive, such as the one used in this item, are more frequently accepted than genetic ones that are based on quantitative restrictions. Due to the close linkages between these units that are essential for their survival, some researchers characterize barrier islands as structures composed of many large depositional units.

Barrier Reef

A barrier reef is defined as a reef separated from the main shore by a deep waterway or lagoon. The initial creation takes place in open water off the coast, with continuing expansion parallel to or toward the shoreline. Barrier reefs with shallow lagoons can resemble fringing reefs in appearance, but they differ due to their genesis and proclivity towards deeper lagoons. Barrier reefs often grow in open water rather than immediately close to a shoreline, and they only become associated with a shoreline due to the evolution of the substrate surrounding the reef or a shift in sea level (Schmitz, 2012). Barrier reefs are uncommon, although they include some of the world's most well-known reefs, such as the Great Barrier Reef of Australia and Belize's Barrier Reef. The Great Barrier Reef in Australia was built on a coastal plain flooded with a fast rise in sea level between 20,000 and 6,000 years ago. The flooded hills of the coastal plain provided shallow regions for corals to grow, and these locations now make-up the reefs and cays of the Great Barrier Reef.

Basal

Basal or basilar is a term meaning base, bottom, or minimum. The act of a glacier sliding across the bed due to meltwater under the ice acting as a lubricant is known as basal sliding. The temperature of the area, the glacier's slope, the bed roughness, the volume of meltwater from the glacier, and the glacier's size all have a role in its movement.

The movement of these glaciers as they slide is jerky, and any seismic activity, especially at the glacier's base, might trigger movement. The majority of the movement is due to pressurized meltwater or extremely tiny water-saturated sediments beneath the glacier (Rao, 2012). This provides a much smoother surface for the glacier to slide on, as opposed to a rough surface that tends to slow the sliding speed. Although meltwater is the most common source of basal sliding, it has been shown that water-saturated sediment can also play up to 90% of the basal movement these glaciers make.

Batholith

Batholiths are igneous rocks formed when magma rises through the earth's crust but does not explode onto the surfacc. The lava cools beneath the surface of the earth, building a rock structure that spans at least 100 square kilometers (40 square miles) and extends to an unknown depth (Figure 19).

Figure 19. Yosemite.

Source: https://en.wikipedia.org/wiki/File:Yosemite_20_bg_090404.jpg.

Bedrock

The firm, solid rock beneath surface materials such as dirt and gravel are known as bedrock. On the ocean floor, bedrock also supports sand and other deposits. Bedrock is a type of cemented rock that is solid and strongly linked together. Unconsolidated rock, which is made composed of loose particles, is frequently found on top of the surface.

Bedrock can stretch hundreds of meters beneath the Earth's surface, all the way to the Earth's crust's base. The rockhead is the highest edge of bedrock (Rivet and Kastens, 2012).

Bedrock may be overlain with saprolite above the rockhead. Saprolite is a type of bedrock that has been heavily weathered or worn away. Chemical weathering has actually taken place on saprolite. This means that saprolite is more than just less-consolidated bedrock; it also has a different chemical makeup. Flowing water or ice has changed the chemical make-up of the bedrock by interacting with minerals.

Layers of soil, sand, or sediment may exist above the saprolite. Typically, these are younger, unconsolidated rocks.

On some mountaintops, along rocky beaches, in stone quarries, and on plateaus, exposed bedrock can be observed. Outcroppings or outcrops are common names for these visible bedrock exposures. Natural processes such as erosion and tectonic uplift can expose outcrops. Drilling can also be used to access outcroppings.

Big Bang

The Big Bang theory describes how the universe began and evolved into the stars and galaxies we see today. From the beginning of time to the present, scientists have referred to the most accepted theory of the universe as the Big Bang (Pyle et al., 2018).

There were no stars, atoms, shape, or structure when the cosmos began as a very hot, tiny, and dense superforce (a blend of the four fundamental forces) (called a "singularity"). Then, around 13.8 billion years ago, space expanded at a breakneck speed (thus the name "Big Bang"). This triggered the formation of atoms, which in turn triggered the formation of stars and galaxies. In 1927, Georges was the first to notice that an expanding cosmos could be traced back in time to a single point of genesis. Today, the cosmos is still expanding and becoming colder.

As a whole, the universe is growing and the temperature is falling as time passes. Cosmology is the study of how the universe began and its development. Some scientists who study cosmology have agreed that the Big Bang theory matches what they have observed so far.

The Big Bang theory is based on a variety of findings. The redshift of very far away galaxies is the most essential. The Doppler effect in light is known as redshift. Because the wavelength of light emitted by an object stretches as it moves away from Earth, its color rays appear to be more similar to the color red than they are. Because red is the longest wavelength in the visible spectrum, scientists use the term "red hot" to characterize this stretched light wave (Press, 2008). The greater the redshift, the faster the item moves away. Scientists were able to verify that the universe is expanding by measuring the redshift, and they were able to calculate how fast the object is traveling away from the Earth. Scientists believe the cosmos was a singularity roughly 13.8 billion years ago, based on very precise observations and measurements.

Because most things become cooler as they expand, scientists believe the cosmos began as a small and hot place.

The abundance of chemical elements in the universe is another evidence that supports the Big Bang idea. Quantities of very light elements like hydrogen, helium, and lithium appear to support the Big Bang idea. "Cosmic microwave background radiation" has also been discovered by scientists. Radio waves are a type of radiation that can be found all around the universe. This radiation is today very feeble and chilly, but it was formerly incredibly powerful and hot.

Biome

A biome is a classification system for areas based on the species that live there. Temperature ranges, soil types, and light and water availability are all unique to a location and serve as niches for specific species, allowing scientists to identify the biome. Scientists, on the other hand, disagree over the number of biomes that exist (Perdigão et al., 2020). Some list six (forest, grassland, freshwater, marine, desert, and tundra), while others count eight (separating two types of woods and adding tropical savannah), while still others are more particular and count as many as 11 (forest, grassland, freshwater, marine, desert, and tundra).

Biomass is the weight or total quantity of living creatures in a single animal or plant species (species biomass) or across all species in a community (community biomass), and is usually referred to as a unit area or volume of habitat. The standing crop is the weight or number of organisms in a region at any given time. The total amount of organic material generated by living creatures in a certain area over a specific time period, known as primary or secondary productivity (the former for plants, the latter for animals), is commonly measured in energy units such gram calories or kilojoules per square meter per year. Weight measurements, such as tons of carbon per square kilometer per year or gigatons of carbon per year, are also used.

In a distinct but related sense, biomass refers to plant and animal waste that is used primarily as a source of energy.

Bioremediation

Bioremediation is a field of biotechnology that involves the removal of contaminants, pollutants, and toxins from soil, water, and other environments using living organisms such as microorganisms and bacteria. Bioremediation is a technique that can be used to clean up contaminated groundwater or environmental issues like oil spills (Parkes and Morris, 2001).

Biosphere

The biosphere is made up of all the places on the planet where life can be found. From the deepest root systems of plants to the gloomy environment of ocean trenches, lush rain forests, and high mountaintops, the biosphere encompasses everything.

Scientists use spheres to describe the Earth. The lithosphere is the Earth's solid surface layer. Above the lithosphere is the atmosphere, which is a layer of air. The hydrosphere is made up of the water on the Earth's surface, in the ground, and in the air.

The biosphere encompasses all three domains since life exists on the land, in the air, and in the water. Although the biosphere spans around 20 kilometers (12 miles) from top to bottom, practically all species may be found between 500 meters (1,640 feet) below sea level and 6 kilometers (3.75 miles) above sea level (Özyurt et al., 2021).

For nearly 3.5 billion years, the biosphere has existed. Prokaryotes, the earliest life forms in the biosphere, were able to thrive without oxygen. Single-celled creatures like bacteria and archaea were once considered prokaryotes.

A chemical reaction developed by some prokaryotes. They were able to use sunlight to convert water and carbon dioxide (CO_2) into simple sugars and oxygen, a process known as photosynthesis. Because these photosynthetic creatures were so numerous, they altered the ecosystem. The atmosphere evolved a mix of oxygen and other gases that may support new forms of life over time.

Bitumen

Bitumen, commonly known as asphaltum or tar, is a black, oily, viscous type of petroleum that forms naturally as plants decompose. This extraordinary natural substance, which is waterproof and combustible, has been used by humans for at least 40,000 years for a range of chores and instruments. In today's world, there are a variety of processed varieties of bitumen used for paving streets and roofing houses, as well as additives to diesel or other gas oils (Orion, 2019).

Bitumen is made up of a complicated hydrocarbon mixture that includes calcium, iron, sulfur, and oxygen. The quality of the substance and the simplicity with which it may be produced is determined by the source and type of crude oil used.

Bitumen is mostly used in the construction industry. It was first employed for its natural adhesive and waterproofing properties, which helped to bind building materials together and lining the bottoms of ships. It has also been used in the past as a medicine.

Black Hole

A black hole is a region of space in which gravity is so strong that even light cannot escape. Because stuff has been crammed into such a small space, gravity is extremely powerful. When a star is dying, this can happen.

People cannot perceive black holes because no light can escape. They are undetectable. Space telescopes equipped with specialized equipment can aid in the discovery of black holes. Special instruments can see how stars in close proximity to black holes behave differently from other stars (Oreskes, 2015).

Black holes can be very large or very small. The smallest black holes, according to scientists, are as small as one atom. These black holes are really small, but they have the bulk of a mountain. The amount of matter, or "stuff," in a thing is measured in mass.

"Stellar" black holes are a different type of black hole. Its mass can be up to 20 times greater than that of the Earth.

In Earth's galaxy, there could be a large number of star mass black holes. The Milky Way is the name of our galaxy.

"Supermassive" black holes are the largest. These black holes have a combined mass of more than a million suns. Scientists have discovered confirmation that at the heart of every giant galaxy is a supermassive black hole. Sagittarius A is the name of the supermassive black hole in the center of the Milky Way galaxy. It has the mass of around 4 million suns and would fit inside a massive ball large enough to hold a few million Earths (Oreskes, 2013).

The smallest black holes, according to scientists, formed when the universe began. Stellar black holes are formed when the core of a massive star collapses in on itself. When this happens, a supernova occurs. A supernova is a star that explodes and sends a portion of its mass into space. Scientists think supermassive black holes were made at the same time as the galaxy they are in.

Black Ice

When water hits a cold road and quickly freezes, a thin, clear layer of ice called black ice forms. It is called black ice because unlike snow, which is white, black ice is transparent, revealing the black road below.

Blackbody

In physics, a blackbody, is a surface that absorbs all radiant radiation that strikes it. The word comes from the fact that incident visible light is absorbed rather than reflected, making the surface appear black. The concept of such a perfect energy absorber is particularly important in the study of radiation phenomena, such as Planck's radiation equation for the spectral energy distribution of reemitted radiation after it has been absorbed (Odenwald, 2020).

A small hole in a box with a blackened interior is the finest practical blackbody since virtually no radiation entering such a hole could escape and would be absorbed inside. A surface covered in lampblack absorbs around 97% of incident light and can be regarded a blackbody for most applications. Polished metal surfaces, on the other hand, absorb only around 6% of incident radiation and reflect the remainder.

Blizzard

Most people think of a blizzard as a bad snowstorm, but a winter storm must meet certain criteria to be classified as a blizzard. According to the National Weather Service, a blizzard is a weather event that includes low temperatures, wind speeds greater than 56 kilometers (35 miles) per hour, and a large amount of falling or blowing snow that lowers visibility to 0.4 kilometers (0.25 mile) for a minimum of three hours (Neeck and Volz, 2011). These whiteout conditions can cause car accidents and people on foot to become lost. Additionally, the colder temperatures that often follow a blizzard can put people at risk of frostbite or hypothermia (Figure 20).

Figure 20. Heavy snow during the January 2016 United States blizzard.

Source: https://en.wikipedia.org/wiki/Blizzard#/media/File:Blizzard3_-_NOAA.jpg.

Blowout

Blowouts are sandy depressions formed by wind removal of sediments in a sand dune environment (psammosere).

Blowing outs happen when wind erodes pockets of bare sand on stabilized vegetated dunes. They are common along the shore and on the outskirts of arid areas. Blowing out does not usually occur on actively flowing dunes because the dunes must be bound in some way, such as by plant roots.

Because of the greater desiccation and disturbances that occur on the higher areas of stabilized dunes, which allows for more surface drag and sediment entrainment when the sand is bare, these depressions frequently originate there (Movva et al., 2012). Most exposed regions quickly re-vegetate before becoming blowouts and expanding; but, when conditions are favorable, wind erosion can scrape the exposed surface and create a tunneling effect, increasing local wind speed. A depression may then develop until it hits a non-erodible substrate, or morphology limits it. The eroded substances climb the steep slopes of the depression and become deposited on the downwind side of the blowout which can form a dune that covers vegetation and lead to a larger depression; a process that helps create parabolic dunes.

Blue Star

Blue stars are composed of the same elements as all other stars in the Universe: around 75% hydrogen, 24% helium, and trace amounts of other elements. So, what makes a blue star blue in the first place?

The temperature of a star determines its hue. The stars that are the coolest appear red, while the hottest look blue. And for a star, its mass is the only factor that determines its temperature. Blue stars have a mass of at least 3 times that of the Sun and can be as massive as the Sun. A star that is 10 times the mass of the Sun or 150 times the mass of the Sun will seem blue to human eyes (Müller et al., 2019).

Nuclear fusion events at the core of stars provides energy. The growing heat of the reactive core and the gravitational forces that hold the star together maintain a dynamic equilibrium throughout the star's life. Fusion produces extremely high energy. Fusion releases some of the energy that binds the particles of the nucleus together, unleashing remarkable power.

Boil

When a liquid is heated to its boiling point, it converts into a vapor, which is known as boiling. When the vapor pressure of a liquid equals the atmospheric pressure exerted on the liquid; and the liquid transitions to a gaseous phase. Boiling is a physical transformation that does not change the chemical makeup of the molecules.

Bubbles emerge when atoms or molecules in a liquid are able to spread out enough to transition from a liquid to a gaseous state.

Bolide

A bolide is typically thought to refer to a very bright meteor, however the phrase can be defined in a variety of ways depending on the context. Any huge crater-forming body, especially one that bursts in the atmosphere, could be referred to by this term. It is a common misnomer for a fireball; however, it is usually reserved for those with an apparent magnitude of 14 or brighter (Metz, 2008).

A bolide is defined as a fireball with an apparent magnitude of 14 or higher, which is more than twice as luminous as the full moon.

A bolide, according to another definition, is any huge crater-forming impacting body whose composition is unknown (for example, whether it is a stony or metallic asteroid or an icy comet).

A superbolide is a bolide that shines brighter than the full moon, with an apparent magnitude of 17 or higher. The Sutter's Mill meteorite in California is a recent example of superbolides.

Boulder

A boulder is a rock piece that is more than 256 millimeters (10.1 in) in diameter in geology. Cobbles and pebbles are the smaller fragments. Some boulders are tiny enough to move or roll by hand, while others are gigantic. A boulder is a huge object that is difficult for a person to move (Li and Li, 2016). Smaller boulders are commonly referred to as rocks or stones in American English (In British English a rock is larger than a boulder). The word boulder comes from the Middle English Balderston or the Swedish Bullersten, which means boulder stone.

Glacial erratics are abundant in areas covered by ice sheets during Ice Ages, such as Scandinavia, northern North America, and Siberia. Boulders scooped up by ice sheets during their advance and deposited as they melt are known as erratics (Schmitz, 2011). Because they are often of a different rock type than the bedrock on which they are deposited, they are dubbed "erratic." One such boulder is used as the pedestal of the Bronze Horseman in Saint Petersburg, Russia.

Breccia

Breccia is a lithified sedimentary rock made up of angular or subangular particles that are greater than 2 millimeters in size (0.08 inch). It is not the same as a conglomerate, which is made up of rounded clasts.

Following that is a brief discussion on breccias. See Sedimentary Rock: Conglomerates and Breccias for more information.

Breccia is divided into three types: sedimentary, pyroclastic, or igneous, and cataclastic. The constituent fragments may be of a single lithology, such as a rock broken in place, or a mixture of lithologies, implying that they accumulated from previously deposited rocks. Cementing materials made of carbonate, silica, or silt can be used to fill void areas between these clasts (Ledley et al., 2011).

Brittle

Brittleness is a quality of a material that cracks when stressed yet has a small tendency to distort before breaking. Brittle materials have a limited deformation capacity, a low capacity to resist impact and vibration, a high compressive strength, and a low tensile strength. The majority of non-metallic inorganic materials are fragile.

Butte

Buttes are rock towers that are tall, flat-topped, and steep-sided. Buttes are formed via erosion, which is the progressive wearing away of the Earth by water, wind, and ice. Buttes were previously part of mesas or plateaus, which were flat, high expanses of land. The sole distinction between a mesa and a butte is their size. A butte, according to most geographers, is taller than it is wide, but a mesa is a considerably bigger, somewhat lower-elevated feature.

Buttes are formed when streams cut through a mesa or plateau over time. Caprock, the hard top layer of buttes, resists weathering and erosion. As a result, the structures are roughly the same height as the plateau or mesa that formed them (Lewis, 2008).

The weaker rock around the caprock is slowly eroded by weathering and erosion, which is most typically caused by wind and moisture. The more susceptible rock beneath caprock is protected by it. Buttes gradually morph into thin spires. Even caprock eventually succumbs to extreme weathering and erosion. Scree or talus refers to the debris that falls to the sides of buttes.

Buttes are most common in desert areas, such as Mexico and the southwest United States. Monument Valley, located in the US states of Utah and Arizona, is home to the world's most famous collection of buttes (Košler, 2007). From Forrest Gump to Easy Rider, the buttes of Monument Valley have served as the backdrop for a slew of films and television shows.

C

Calcium

Calcium is an atomic number 20 chemical element with the symbol Ca. Calcium is a reactive alkaline Earth metal that develops a black oxide-nitride layer when exposed to air. It shares the same physical and chemical properties as its heavier homologs strontium and barium. After iron and aluminum, it is the fifth most plentiful element in the Earth's crust and the third most abundant metal (Kleinhans et al., 2010). Calcium carbonate is the most prevalent calcium compound on Earth, and it may be found in limestone and the fossilized remains of early sea life. Other calcium compounds include gypsum, anhydrite, fluorite, and apatite. The name comes from the Latin word calx, which means "lime," and refers to the product of heating limestone.

Calcium Carbonate

The chemical compound calcium carbonate has the formula $CaCO_3$. It is the principal component of eggshells, snail shells, seashells, and pearls and is found in rocks as the minerals calcite and aragonite (most famously as limestone, which is a form of sedimentary rock consisting primarily of calcite). The active element in agricultural lime is calcium carbonate, which is formed when calcium ions in hard water combine with carbonate ions to form limescale. It's used in medicine as a calcium supplement or an antacid, but too much of it can be dangerous, causing hypercalcemia and intestinal problems (Klippel et al., 2019).

Caldera

A caldera is a vast cauldron-like hollow formed when a magma chamber is emptied after a volcanic eruption. When enormous amounts of magma are erupted in a short period of time, the rock above the magma chamber loses structural support. The ground surface then descends into the magma chamber, which has been emptied or partially empty, leaving a large depression at the surface (from one to dozens of kilometers in diameter). The feature is commonly referred to as a crater, but it is actually a sinkhole because it is generated by subsidence and collapse rather than an explosion or collision. Caldera development is an uncommon phenomenon, occurring only a few times every century, compared to the thousands of volcanic eruptions that occur each century. Between 1911 and 2016, only seven caldera-forming collapses are known to have occurred. At the year 2018, a caldera collapsed in Klauea, Hawaii.

Caldera is derived from the Spanish caldera and the Latin caldaria, which both mean "cooking pot." The English phrase cauldron is also used in some writings, albeit it now refers to a caldera that has been thoroughly eroded to expose the beds beneath the caldera floor in more recent study (Hundman and Mattmann, 2017). When the German geologist Leopold von Buch wrote his memoirs of his 1815 voyage to the Canary Islands, where he first witnessed the Las Caadas crater on Tenerife, with Mount Teide dominating the landscape, and then the Caldera de Taburiente on La Palma, he coined the term caldera.

Calorie

The calorie is an energy unit defined as the amount of heat required to raise the temperature of one kilogram of water by 1°C (or kelvin).

Two primary definitions of calorie are widely used for historical reasons. The amount of heat required to raise the temperature of one gram of water by 1°C is known as the tiny calorie or gram calorie (or kelvin). The amount of heat required to cause the same increase in one kilogram of water is measured in the large calorie, food calorie, or kilocalorie (Cal, Calorie, or kcal), which is most commonly employed in nutrition (Guo et al., 2016). As a result, one kilocalorie (kcal) equals 1,000 calories (cal). Calorie is a term used in food science to refer to a huge calorie (with a capital C by some authors to distinguish from the smaller unit). In most countries, labels of industrialized food products are required to indicate the nutritional energy value in (kilo or large) calories per serving or per weight.

Calorie relates directly to the metric system, and therefore to the SI system. It has been regarded as obsolete within the scientific community since the adoption of the SI system, but is still in some use. The SI unit of energy is the joule. One calorie is defined as exactly 4.184 J, and one Calorie (kilocalorie) is 4,184 J.

Calving

The breaking of ice chunks from the edge of a glacier is known as ice calving, glacier calving, or iceberg calving. It is an ice ablation or ice disturbance technique. A mass of ice from a glacier, iceberg, ice front, ice shelf, or crevasse is suddenly released and breaks away. It is possible that the iceberg that breaks away is an iceberg, but it might also be a growler, bergy bit, or crevasse wall breakaway.

Before slabs of ice up to 60 meters (200 feet) high break loose and tumble into the water, glaciers typically make a loud cracking or booming sound (Kleinhans et al., 2010). When ice melts into the water, it creates enormous, potentially dangerous waves. The waves that erupt at places like Johns Hopkins Glacier can be so big that vessels could not get closer than 3 kilomcters away (1.9 mi) (Goldstein et al., 2017).

Capillary Action

Capillary action (also known as capillary motion, capillary effect, or wicking) is the process of a liquid flowing in a tiny space without the aid of, or even in opposition to, other forces such as gravity. Liquids can be drawn up between the hairs of a paintbrush, in a narrow tube, in porous materials like paper and plaster, in non-porous materials like sand and liquefied carbon fiber, or in a biological cell. Intermolecular forces between the liquid and the surrounding solid surfaces create this. If the tube's diameter is small enough, the liquid is propelled by a combination of surface tension (produced by cohesion within the liquid) and adhesion forces between the liquid and the container wall.

Carbon

Carbon (from Latin: carbo "coal") is an atomic number 6 chemical element with the symbol C. It is nonmetallic and tetravalent, which means it has four electrons available for chemical bonding. It belongs to the periodic table's 14th group. Only roughly 0.025% of the Earth's crust is made up of carbon (Escorihuela, 2018). The stable isotopes 12C and 13C are found in nature, but 14C is a radioactive with a half-life of around 5,730 years. Carbon is one of the few elements that has been known since the beginning of time (Figure 21).

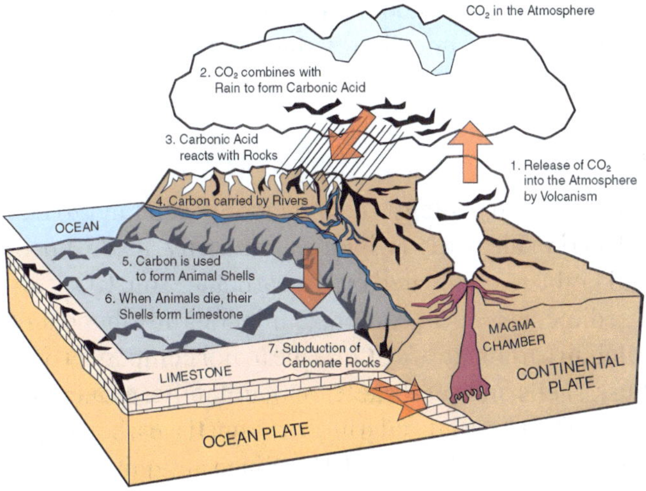

Figure 21. The origin atmosphere of the Earth was rich in reduced gases including methane, CH_4. The carbon content of the Earth steadily increased over eons as a result of collisions with carbon-rich meteors. As the oxygen content of the atmosphere increase, the carbon-containing molecules were oxidized to CO_2.

Source: http://butane.chem.uiuc.edu/pshapley/Environmental/L29/2.html.

After hydrogen, helium, and oxygen, carbon is the 15th most prevalent element in the Earth's crust and the fourth most plentiful element in the universe by mass. Carbon's abundance, unique range of organic compounds, and exceptional ability to form polymers at temperatures routinely found on Earth make it a common component of all known life.

After oxygen, it is the second most prevalent element in the human body in terms of mass (approximately 18.5%).

Carbon atoms can link in a variety of ways, resulting in a variety of carbon allotropes. Graphite, diamond, amorphous carbon, and fullerenes are examples of well-known allotropes. The allotropic form of carbon has a wide range of physical properties. Graphite, for example, is opaque and black, but diamond is translucent. Graphite is soft enough to leave a streak on paper (thus its name, which comes from the Greek verb "v" meaning "to write"), but diamond is the hardest naturally occurring mineral (Duffy, 2014). Diamond has a poor electrical conductivity, but graphite has a high electrical conductivity. Diamond, carbon nanotubes, and graphene have the highest thermal conductivities of all known materials under normal conditions. Under normal conditions, all carbon allotropes are solids, with graphite being the most thermodynamically stable at standard temperature and pressure. They are chemically resistant and require high temperature to react even with oxygen.

Carbonate

The term "carbonate" in geology and mineralogy can refer to both carbonate minerals and carbonate rock (which is primarily made up of carbonate minerals), and both are dominated by the carbonate ion, CO_3^{2-}. In chemically precipitated sedimentary rock, carbonate minerals are abundant and diverse. Calcite, or calcium carbonate, $CaCO_3$, is the major component of limestone (as well as mollusk shells and coral skeletons); dolomite, a calcium-magnesium carbonate $CaMg(CO_3)_2$; and siderite, or iron(II) carbonate, $FeCO_3$, is a significant iron resource (Clowes, 2010). Since antiquity, sodium carbonate ("soda" or "natron") and potassium carbonate ("potash") have been used for cleaning, preservation, and the production of glass. Carbonates are widely utilized in industry, including in iron smelting, as a raw ingredient in the making of Portland cement and lime, and in the composition of ceramic glazes, among other things.

Cascade

In geomorphology, a cascade/cascading system is a form of dynamic system characterized by mass and energy transmission along a chain of component subsystems, with the output of one subsystem becoming the input of the next. The valley glacier, for example, receives snowfall and rock debris from the slopes above, as well as potential energy (derived from elevation), which is cascaded through a series of climatic environments with a progressive reduction in mass and dissipation of energy, with the output of the glacier being sediment and water, which are the input to the proglacial subsystem.

Catalyst

In chemistry, a catalyst is any substance that speeds up a reaction without being consumed. Many vital metabolic reactions are catalyzed by enzymes, which are naturally occurring catalysts (Clary and Wandersee, 2011).

Metals and their oxides, sulfides, and halides, as well as the semimetallic elements boron, aluminum, and silicon, make up the majority of solid catalysts. Solid catalysts are usually disseminated in other substances known as catalyst supports; gaseous and liquid catalysts are commonly utilized in their pure form or in combination with suitable carriers or solvents.

Cation

A positively charged ion, i.e., one that would be attracted to the cathode in electrolysis.

Cavern

A cave, sometimes known as a cavern, is a massive natural aperture in the Earth that can be explored by humans. Many different types of rock and processes can result in the formation of such a hole. Caves created by a chemical reaction between circulating groundwater and limestone or dolomite bedrock are the largest and most numerous. Solution caves are a type of cave that is commonly seen in karst terrain (Cossu et al., 2010). Karst topography is characterized by a rugged and jumbled environment of bare bedrock ledges, disturbed surface drainage, sinkholes, and caves, and is named after the Karst region of the western Balkan Peninsula, which stretches from Slovenia to Montenegro. It should be emphasized, however, that karst areas vary greatly. Some may have spectacular surface landforms, yet there are few caves. Others, on the other hand, may have considerable cave development but no visible surface expression for example, the Guadalupe Mountains of New Mexico, the site of Carlsbad Caverns and various other caves, have very few surfaces karst features.

Celestial

The term heavenly body encompasses the entire known and unknown universe. A celestial body is defined as any natural body that exists outside of the Earth's atmosphere. The Moon, Sun, and other planets in our solar system are simple examples. However, those are just a few examples (Cerase, 2019). Many celestial bodies can be found in the Kuiper belt. A celestial body is any asteroid in space. So, with such a big subject, what do you write about? Let us start with a sampling of five of my favorites and leave it at that for the time being. However, which five? Let us take a look at Ceres, the Kuiper belt, Cruithne, Achernar, and Apophis.

Ceres is a celestial body in the belt between Mars and Jupiter that is by far the largest and most massive asteroid. It has a mass of 9.51020 and is roughly the size of Texas (975 km × 909 km).

It makes up a third of the total mass of the asteroid belt. It has enough mass to support self-gravity, which is a prerequisite for being classified as a dwarf planet. It orbits the Sun every 1679.819 days and has a very slight axial tilt. The temperature on the surface is moderate. The high temperature is expected to be in the –38°C range (235 K). The apparent brightness magnitude of Ceres ranges from +6.9 to +9. Ceres is nearly brilliant enough to be seen with the naked eye when it is at its brightest point. On a fully black night, it can be viewed with binoculars whenever it is above the horizon.

Celsius

The degree Celsius is a temperature unit on the Celsius scale, which was first known as the centigrade scale. The degree Celsius (symbol: °C) can refer to a specific temperature on the Celsius scale or a unit used to denote a temperature difference or range. It was named after Anders Celsius (1701–1744), a Swedish astronomer who invented a comparable temperature scale. The measure was previously known as centigrade, from the Latin words centum, which means 100, and gradus, which means steps, until being renamed to honor Anders Celsius in 1948 (Britner, 2008). Since 1743, the Celsius scale has been based on the freezing point of water being 0°C and the boiling point of water being 100°C at 1 atmosphere pressure. The values were inverted prior to 1,743. (i.e., the boiling point was 0° and the freezing point was 100°). Jean-Pierre Christin proposed the 1,743 scale reversal.

Cenozoic

The Cenozoic is the current geological epoch of Earth, covering the last 66 million years of its existence. The dominance of animals, birds, and flowering plants, as well as a cooling and drying environment and the current structure of continents, describe it. It is the most recent of three geological epochs since the emergence of complex life, with the Mesozoic and Paleozoic epochs preceding it (Butler, 2016). It began with the Cretaceous–Paleogene extinction event, when numerous species went extinct, including non-avian dinosaurs, in an event attributed by most experts to the Chicxulub impactor, a massive asteroid or other celestial body.

Chain Reaction

A chain reaction is a sequence of reactions where a reactive product or by-product causes additional reactions to take place. In a chain reaction, positive feedback leads to a self-amplifying chain of events.

Chamber

A magma chamber is a vast pool of liquid rock located beneath the Earth's surface. In such a chamber, the molten rock, or magma, is less dense than the surrounding country rock, causing buoyant forces on the magma to push it upwards. When magma finds a way to the surface, it causes a volcanic explosion; as a result, many volcanoes are built over magma chambers (Bohle, 2015). Because these chambers are difficult to detect deep within the Earth, the majority of those that have been discovered are close to the surface, often between 1 and 10 kilometers below the surface.

Channel

A channel is a large strait or waterway that connects two landmasses that are close in proximity. A channel is a narrow body of water that connects two bigger bodies of water. It can also be the deepest point of a river.

Glaciers cut down deep valleys between two landmasses, forming some of the channels. People generally dig channels from the bottoms of small waterways so that huge ships can pass through. This is referred to as navigation channels.

Chemical

A chemical substance is a type of stuff with a predictable chemical makeup and behavior. Physical separation procedures, i.e., breaking chemical bonds, cannot separate a chemical substance into its constituent elements, according to some references. Simple chemicals, chemical compounds, and alloys are all examples of chemical substances. Depending on expert opinion, chemical elements may or may not be included in the definition (Zhuang et al., 2019).

A chemical bond is a long-term attraction between atoms, ions, or molecules that allows chemical compounds to form. Ionic bonds are formed by the electrostatic force of attraction between oppositely charged ions, while covalent bonds are formed by the sharing of electrons. Chemical bonds come in a variety of strengths; there are "strong bonds" or "primary bonds" like covalent, ionic, and metallic connections, as well as "weak bonds" or "secondary bonds" like dipole–dipole interactions, the London dispersion force, and hydrogen bonding.

A chemist is a scientist who has studied chemistry. Chemists examine the qualities and composition of matter. Chemists meticulously characterize the features they examine in terms of quantities, down to the level of molecules and the atoms that make them up. Chemists quantify substance quantities, chemical reaction rates, and other chemical parameters with great precision. Pharmacists are often referred to as chemists in Commonwealth English.

Chert

Chert is a hard, fine-grained sedimentary rock made up of microcrystalline or cryptocrystalline quartz, which is silicon dioxide in its mineral form (SiO_2). Chert is often biological, but it can also form inorganically as a chemical precipitate or diagenetic replacement, such as in petrified wood (Figure 22).

Figure 22. This specimen of gray chert is about two inches (five centimeters) across. It breaks with a smooth conchoidal fracture. Edges of the piece have sharp edges as a result of the conchoidal fracture.

Source: https://geology.com/rocks/chert.shtml.

Chert is made up of petrified siliceous ooze, a biogenic sediment that covers extensive portions of the deep ocean floor and contains the silicon skeletons of diatoms, silicoflagellates, and radiolarians (Yu and Gong, 2012). The existence of fossil cyanobacteria in Precambrian cherts is significant. Chert occasionally contains macrofossils in addition to microfossils. Some chert, on the other hand, is fossil-free.

Chlorine

Chlorine is a chemical that is utilized in industry and can be found in some consumer items. Chlorine can also be inhaled as a toxic gas. Chlorine gas can be compressed and cooled to become a liquid, which can then be transported and stored. When liquid chlorine is released, it swiftly transforms into a gas that stays near the ground and spreads quickly.

The stench of chlorine gas, which is similar to that of bleach, is harsh and disagreeable. The strong odor may be enough to alert individuals that they are being exposed.

The hue of chlorine gas is yellow-green. Chlorine is not flammable in and of itself, but it can react explosively with other chemicals like turpentine and ammonia to generate explosive compounds.

Chlorofluorocarbon

CFC stands for chlorofluorocarbon, which is an organic molecule made up of carbon, fluorine, and chlorine. Hydrochlorofluorocarbons, or HCFCs, are CFCs that also contain hydrogen in place of one or more chlorines. CFCs are also known as Freons, a trademark of Wilmington, Delaware-based E.I. du Pont de Nemours, and Company. CFCs were first developed in the 1930s as a refrigerant (Witze, 2013). Trichlorofluoromethane (CFC-11) and dichlorodifluoromethane (CFC-12) were used as aerosol-spray propellants, solvents, and foam-blowing agents, respectively. They are ideal for these and other uses since they are harmless and nonflammable, and they can easily transition from liquid to gas and vice versa.

Chromosphere

The chromosphere, or "sphere of color," is the second of the Sun's three major layers, measuring 3,000 to 5,000 kilometers in depth. It only shows its rosy red color during eclipses. The chromosphere is located directly above the photosphere and below the solar transition area. The chromosphere layer on top of the photosphere is homogeneous. From the homogenous layer, a forest of hairy-appearing spicules rises, some of which reach 10,000 kilometers into the corona above.

The chromosphere's density is only 104% that of the photosphere, the layer underneath it, and 108% that of Earth's atmosphere at sea level. The chromosphere is generally invisible, and only visible during a total eclipse, when its reddish tint is revealed. The color palette ranges from pink to crimson. Due to the dazzling brightness of the photosphere beneath, the chromosphere cannot generally be seen without special equipment.

Cirque

Cirques are bowl-shaped, amphitheater-like depressions that glaciers carve into mountains and valley sidewalls at high elevations. Often, the glaciers flow up and over the lip of the cirque as gravity drives them downslope (Wilkinson et al., 2020).

Cirrus

Cirrus clouds are short, detached, hair-like clouds found at high altitudes. These delicate clouds are wispy, with a silky sheen, or look like tufts of hair. In the daytime, they are whiter than any other cloud in the sky. While the Sun is setting or rising, they may take on the colors of the sunset.

Clastic Rock

Clastic rocks are made up of pre-existing mineral and rock bits, or clasts. A clast is a remnant of geological detritus, chunks, and smaller grains of rock that have been physically weathered off of other rocks. Clastic is a term used by geologists to describe sedimentary rocks, particles in sediment transport, whether suspended or as bed load, and sediment deposits.

Cleavage

Cleavage is a sort of planar rock feature that arises as a result of deformation and metamorphism in structural geology and petrology. The sort of cleavage feature that emerges is determined by the degree of deformation and metamorphism, as well as the rock type. These formations are most commonly found in fine-grained rocks made up of minerals that have been altered by pressure solution (Wallace-Wells, 2019).

The tendency of crystalline minerals to split along specified crystallographic structural planes is known as cleavage in mineralogy. The regular positions of atoms and ions in the crystal form smooth repeating surfaces that are visible both in the microscope and to the human eye, resulting in these planes of relative weakness. The crystal will fracture along the weakly bonded planes if the bonds in certain directions are weaker than in others. The term "cleavage" refers to these flat breaks. The classic example of cleavage is mica, which cleaves in a single direction along the basal pinacoid, making the layers seem like pages in a book. In fact, mineralogists often refer to "books of mica."

Climate

The long-term weather pattern in a certain location is referred to as climate.

Hour by hour, day by day, month by month, and even year by year, the weather might change. The climate of a region is determined by its weather patterns, which are normally recorded for at least 30 years.

Climates differ in different places of the world. Every day is hot and wet in some regions of the world. They have a humid, tropical environment. Others are bitterly cold and snow-covered for the most of the year. They live in a polar environment. Many additional climates exist between the freezing poles and the sweltering tropics, all of which contribute to Earth's biodiversity and geologic legacy (Toth et al., 2014).

The climatic system of a region determines the climate. The atmosphere, hydrosphere, cryosphere, land surface, and biosphere are the five major components of a climate system.

The climate system's most variable component is the atmosphere. The composition and flow of gases that surround the Earth can change dramatically as a result of natural and man-made factors.

Changes to the hydrosphere, which include variations in temperature and salinity, occur at much slower rates than changes to the atmosphere.

Coast

The coast is the area of land that is next to or near the ocean. A coastline is a line on a map that indicates the location of a coast, although the term is frequently used to refer to the shore itself. Coastal is an adjective that describes something that is on, near, or linked with the coast.

Because the blend of fresh and salt water in estuaries offers numerous nutrients for marine life, the coast, and its associated areas on and off shore are an essential aspect of a local ecosystem.

Salt marshes and beaches are also home to a diverse range of plants, animals, and insects that are essential to the food chain (Tollefson, 2015).

Coasts, like the water that defines them, are a dynamic ecosystem that is constantly changing. The erosion, accretion, and sculpting of coasts has been caused by natural processes such as sea level rise, waves, as well as flooding and creation of continental shelves and drowned river valleys (rias). and numerous weather occurrences.

Cold Front

A cold front is the leading edge of a cooler mass of air that replaces a warmer mass of air at ground level and lies within a deep trough of low pressure. It forms along the leading edge of an extratropical cyclone's cold air advection pattern—known as the cyclone's dry "conveyor belt" flow—often to the west in the Northern Hemisphere, to the east in the Southern. Temperature differences across the barrier can reach 30°C (54°F) on one side and 30°C (54°F) on the other (Thessen, 2016). Rain can fall along the boundary if there is enough moisture present. A narrow line of thunderstorms can emerge along the frontal zone if there is substantial instability along the boundary. If the front's instability is weak, a large shield of rain might sweep in behind it, increasing the temperature differential across the front due to evaporative cooling. Cold fronts are stronger in the fall and spring transition seasons and weakest during the summer.

Comet

A comet is a tiny, frozen entity in the Solar System that warms up and begins to leak gases as it approaches the Sun, a process known as outgassing. This results in a visible atmosphere or coma, as well as a tail in rare cases. The impact of solar radiation and the solar wind on the nucleus of the comet cause these behaviors. Comet nuclei are made up of loose collections of ice, dust, and small rocky particles that range in size from a few 100 meters to tens of kilometers wide. The coma might be up to 15 times the size of Earth, and the tail could be more than one astronomical unit long. A comet can be observed without a telescope from Earth if it is brilliant enough, and it can cover an arc of 30° (60 Moons) across the sky. Many nations and religions have viewed and recorded comets from ancient times.

Composite

A composite material (also known as a composition material or composite, which is the popular word) is one that is made up of two or more constituent materials. These constituent materials have markedly different chemical or physical properties, and when combined, they form a material having features that are not found in the individual constituents. Individual constituents stay separate and distinct inside the finished construction, distinguishing composites from mixes and solid solutions (Telling et al., 2017).

Compound

A substance that cannot be broken down into a simpler form is referred to as an element. A unique atomic number distinguishes them. In the periodic table, elements are grouped by their atomic number, which highlights atoms with comparable qualities. Compounds are formed when components combine. When two hydrogen atoms bind to an oxygen atom, water is formed. Water is a compound, which is a mixture of two or more elements.

Compression

Compression is a word used in geology to describe a collection of stresses directed toward the core of a rock body. The maximum compressive stress that may be applied to a material before it fails is referred to as compressive strength. Thrust faulting can develop when the maximum compressive stress is horizontal, leading in the shortening and thickening of that area of the crust (Showstack, 2008). A portion of rock will frequently collapse in normal faults when the maximum compressive stress is vertical, horizontally extending and vertically thinning a given layer of rock. Folding of rocks can also be caused by compressive loads. Tectonic-scale deformation is constantly subjected to net compressive stress due to the enormous magnitudes of lithostatic stress in tectonic plates.

Concentration

The degree to which one substance is present in a mixture is referred to as concentration. Each substance's concentration in a mixture can be expressed in mass or volume units. The mixture's components can be gases, liquids, or solids.

The atmosphere of the Earth, for example, contains a mixture of gases, with nitrogen gas accounting for 78% of the overall volume. (Percentages refer to the number of parts of a substance per 100 parts of the mixture.) To put it another way, the nitrogen concentration in the atmosphere is 78% (Schmitz, 2011). Steel is a blend of iron and other metals that comes in a variety of forms.

Concentric

A concentric structure in which approximately parallel layers of a mineral are arranged around a common center.

Concerted

A coordinated reaction is a chemical process in which all bond breaking and bond formation takes place at the same time. There are no reactive intermediates or any high-energy unstable intermediates involved. Because concerted reaction rates are not affected by solvent polarity, a considerable buildup of charge in the transition state is unlikely. Because all bonds are generated and destroyed in unison, the reaction is said to proceed through a coordinated process. Concerted reactions include pericyclic reactions, the SN_2 reaction, and various rearrangements, such as the Claisen rearrangement (Schmitz, 2012).

Condensation

Condensation is the transformation of a gaseous state into a liquid one, and it is the inverse of vaporization. The term is most commonly used to describe the water cycle. It is also known as the transformation of water vapor into liquid water when it comes into contact with a liquid or solid surface, or cloud condensation nuclei in the atmosphere. Deposition occurs when a shift occurs directly from the gaseous to the solid phase.

Conduction

Conduction is one of the three primary methods through which heat energy is transferred from one location to another. Radiation and convection are the other two methods of heat transfer. Heat energy is transferred through collisions between surrounding atoms or molecules in the process of conduction (Rao, 2012).

Solids and liquids, where the particles are closer together, conduct more readily than gases, where the particles are more apart. When there is a big temperature differential between the substances in contact, the rate of energy transfer by conduction is faster.

Cone

Different types of eruptions deposit different materials around the slopes of a volcano, resulting in these formations. Layers are formed as volcanic ash and lava explosions alternate. These layers accumulate over time. As a result, the cone has a slope that is milder than a cinder cone but steeper than a shield volcano (Rivet and Kastens, 2012).

They are made up of congealed lava particles and blobs released from a single vent. The gas-charged lava is violently pushed into the air, breaking into minute particles that harden and fall as cinders surrounding the vent, forming a circular or oval cone.

Confining

The force that is applied to an object is known as stress. Stress is defined as the force applied to a rock per unit area in geology. Materials are subjected to four different types of stresses.

The weight of all the material atop a deeply buried rock pushes it down. The rock cannot deform since it cannot move. Confining stress is the term for this (Pyle et al., 2018).

Compression causes rocks to fold or fracture as they are squeezed together. The most frequent stress at convergent plate borders is compression.

Constellation

In astronomy, a constellation is a grouping of stars that were thought to make prominent configurations of objects or creatures in the sky, at least by those who named them. Astronomers and navigators can utilize constellations to help them find certain stars.

Continent

A continent is a huge landmass made up of multiple smaller landmasses. Up to seven geographical regions are typically regarded as continents, and are generally identified by convention rather than any rigorous criterion. Asia, Africa, North America, South America, Antarctica, Europe, and Australia are the seven regions, in order of area, from largest to smallest. Some of them may be merged in variations with fewer continents; for example, some systems treat Afro-Eurasia, America, and Eurasia as single continents.

To classify all of the world's territory into geographical areas, oceanic islands are typically grouped with a nearby continent. The majority of the Pacific Ocean's Island countries and territories, as well as the continent of Australia, are placed together in this scheme to constitute the Oceania geographical region (Press, 2008).

Continental drift is one of the first theories on how continents moved over time, according to geologists. Wegener released a paper in the early 20th century describing his notion that continental landmasses were "drifting" across the Earth, occasionally plowing through oceans and colliding with one another.

The undersea boundary of a continent is known as a continental shelf. The seven major land divisions on Earth are known as continents. A continental shelf stretches from a continent's coastline to a point known as the shelf break (Plotnick et al., 2014). "The slope seaward from the shelf to the upper edge of a continental rise or the point where there is a general reduction of slope," according to the IHO.

Contour

A contour line is a map line that depicts an imaginary line on the ground surface with all points at the same height above a datum plane, usually mean sea level. Contour lines on a map contour line show relief by joining points of equal elevation, as shown in the diagram.

Control Rod

Nuclear reactors employ control rods to regulate the rate of fission of the nuclear fuel – uranium or plutonium. Chemical elements capable of absorbing numerous neutrons without fissioning, such as boron, cadmium, silver, hafnium, or indium, are present in their compositions (Pennington et al., 2020). For neutrons of differing energy, these elements have distinct neutron capture cross sections. Thermal neutrons are used in boiling water reactors (BWR), pressurized water reactors (PWR), and heavy-water reactors (HWR), whereas fast neutrons are used in breeder reactors. Depending on the energy spectrum of the neutrons, each reactor design can use different control rod materials.

Convection

Weather is influenced by convection. The Sun heats the earth's surface, which then warms and rises as cooler air comes into touch with it, creating an upward circulation in the atmosphere. Wind, clouds, and other weather might result from this current. Convection currents move magma layers within the earth, while currents are created by convection in the ocean. Convection ovens circulate heat via fans, allowing whatever you are cooking to cook faster and more evenly.

Convergence

When plates move towards one other and collide, this is known as convergent (colliding). When a continental plate collides with an oceanic plate, the oceanic plate dips beneath the thicker, more rigid continental plate because it is thinner, denser, and more flexible. This is referred to as subduction (Perdigão et al., 2020). Deep ocean trenches form as a result of subduction, such as the one off the west coast of South America. The continent's dragged-down rocks begin to dissolve. Sometimes molten rock comes to the surface and forms a line of volcanoes throughout the continent. Around 80% of earthquakes happen along convergent borders, when plates are forced together.

Convex

Being convex means being curled outwards. The term "convex" is defined in this way, and it is pretty clear what it means. Like our pizza in the first example, it must be bent outwards on both sides. To be convex, we must be able to draw two points anywhere within the form, and the line connecting them must not travel outside of the shape. The form is not convex if we can connect two places within it and the line joining them passes outside the shape.

Core

The Earth's core is the planet's extremely hot and dense center. Underneath the chilly, brittle crust and the mostly-solid mantle is a ball-shaped core. The core is located approximately 2,900 kilometers (1,802 miles) below the Earth's surface and has a radius of approximately 3,485 kilometers (2,165 miles) (Parkes and Morris, 2001).

The core of the Earth is older than the planet itself. Earth was a homogeneous ball of hot rock when it created roughly 4.5 billion years ago. The ball became considerably hotter as a result of radioactive decay and remaining heat during planetary formation (the collision, accretion, and compression of space rocks). After roughly 500 million years, the temperature of our young planet reached the melting point of iron, which is about 1,538°C (2,800°F). The iron catastrophe is the name given to this pivotal moment in Earth's history. The iron catastrophe accelerated the migration of molten, rocky material on Earth.

Coriolis Effect

The Coriolis effect depicts the deflection pattern of objects that are not firmly linked to the ground when they travel large distances around Earth. Many large-scale weather patterns are caused by the Coriolis effect.

The rotation of the Earth is the key to the Coriolis effect. At the Equator, the Earth rotates faster than it does at the poles. Because the Earth is wider around the Equator, it takes approximately 1,600 kilometers (1,000 miles) each hour to complete a 24-hour rotation. Earth rotates at a slow 0.00008 kilometers (0.00005 miles) per hour near the poles (Özyurt et al., 2021).

Assume you are standing on the Equator and want to toss a ball to a friend in the middle of North America. Because he is traveling slower and has not caught up, the ball will appear to land to the right of your companion if you throw it in a straight line.

Corona

Corona is the plasma-rich outermost part of the Sun's atmosphere (hot ionized gas). It has an extraordinarily low density and a temperature of about 2 million kelvins. The corona's size and shape change throughout time as it is influenced by the Sun's magnetic field. The solar wind is created by the expansion of coronal gases and only ceases at the heliopause. It flows radially outward through the entire solar system.

Cosmic

Relating to the universe or cosmos, especially as distinct from the earth, e.g., "cosmic matter."

Crater

A crater is a bowl-shaped depression or hollowed-out area caused by a meteorite impact, volcanic activity, or an explosion.

Impact craters are craters created when a meteorite collides with the Earth (or another planet or moon). A large meteorite's high-speed impact compresses, or forces downward, a large area of rock. The rock is pulverized by the pressure. The pulverized rock, on the other hand, rebounds almost soon after the blow (Orion, 2019). Massive volumes of shattered debris shoot skyward, leaving a huge, circular hole in the place where the rock used to be. The majority of the material falls around the freshly formed crater's rim.

Craters are frequently formed by volcanic activity. Some volcanic craters have steep walls and are deep. Others are shallow and wide.

A caldera is not the same as a crater. A volcano's outward ejection of rocks and other elements creates craters. Calderas are generated when the magma chamber of a volcano collapses inward. Calderas are sometimes referred to as big craters because craters are normally much smaller than calderas.

An explosion creates a third sort of crater. When materials or chemicals explode, they displace everything in their path. The debris frequently falls in a circular fashion around the explosion site, forming a crater.

Explosions can be either natural or man-made. When water meets with superhot magma from a volcano, for example, an explosion called a maar occurs naturally. Maars are a volcanic crater and a sort of explosive crater.

Craton

The stable interior portion of a continent made up of old crystalline basement rock is known as a craton. The name craton is used to differentiate these areas from movable geosynclinal troughs, which are linear bands of sediment accumulations that are vulnerable to subsidence (i.e., downwarping). Shields and platforms may be found in the vast core cratons of continents (Oreskes, 2015). The section of a craton where (typically) Precambrian basement rocks crop out significantly at the surface is called a shield. In a platform, on the other hand, the basement is covered by horizontal or sub-horizontal sediments.

Cretaceous

The Cretaceous Period was a geological epoch that lasted roughly 145 million years and ended about 66 million years ago (Mya). It is the Mesozoic Era's third and final phase, as well as the longest. It is the longest geological period in the Phanerozoic, lasting roughly 79 million years. The name comes from the Latin word creta, which means "chalk," which was plentiful in the second half of the time. Its German version, Kreide, is typically abbreviated as K.

The Cretaceous was characterized by a warm climate, which resulted in high eustatic sea levels and the formation of numerous shallow inland seas. On land, dinosaurs continued to dominate, while the oceans and seas were occupied by now-extinct marine reptiles, ammonites, and rudists. There was no ice on the planet, and woods reached all the way to the poles. During this epoch, new mammalian and avian species appeared.

Flowering plants arose during the Early Cretaceous and quickly diversified, eventually becoming the dominating group of plants on the planet by the end of the Cretaceous, coinciding with the decline and extinction of previously prevalent gymnosperm taxa (Oreskes, 2013).

The Cretaceous period (together with the Mesozoic) came to an end with the Cretaceous–Paleogene extinction event, which saw the extinction of many groups, including non-avian dinosaurs, pterosaurs, and huge marine reptiles. The abrupt Cretaceous–Paleogene boundary (K–Pg boundary), a geologic hallmark linked with the mass extinction that occurred between the Mesozoic and Cenozoic Eras, marks the end of the Cretaceous.

Crevasse

A crevasse is a deep fracture, crevice, or fissure in ice, glaciers, or the soil. Crevasses arise as a result of the movement and stress caused by shear stress generated when two semi-rigid parts atop a plastic substrate move at different speeds. Breakage occurs along the faces as a result of the shear stress's strength.

Vertical or near-vertical walls are common in crevasses, which can melt and form seracs, arches, and other ice formations. The stratigraphy of the glacier is sometimes exposed by these walls. The size of a crevasse is frequently determined by the amount of liquid water present in the glacier. A crevasse can be up to 45 meters deep and 20 meters wide (Odenwald, 2020).

A snow bridge formed of past years' accumulation and snow drifts may cover, but not necessarily fill, a crevasse.

As a result, crevasses are rendered invisible, and anyone attempting to negotiate their way across a glacier risks death. A snow bridge over an ancient chasm may droop from time to time, providing some landscape relief, but this should not be relied upon. Groups intending to go on a glacier with hidden crevasses must be trained and equipped to maneuver as a rope team.

Crystalline

A crystal, also known as a crystalline solid, is a solid whose constituents (such as atoms, molecules, or ions) are arranged in a highly ordered microscopic structure that forms a crystal lattice that extends in all directions. Furthermore, macroscopic single crystals are usually distinguished by their geometrical shape, which consists of flat sides with distinct orientations (Neeck and Volz, 2011). Crystallography is the scientific study of crystals and crystal formation. Crystallization or solidification is the process of forming crystals through mechanisms of crystal growth.

Snowflakes, diamonds, and table salt are examples of big crystals. The majority of inorganic solids are polycrystals, which are made up of numerous small crystals fused together to form a single solid. Polycrystals can be found in most metals, minerals, ceramics, and ice. Amorphous solids are a third type of solid in which the atoms have no periodic structure. Glass, wax, and many polymers are examples of amorphous solids.

Cumulonimbus

Cumulonimbus (from Latin cumulus, meaning "heaped" and nimbus, meaning "rainstorm") is a dense, towering vertical cloud formed by water vapor propelled by strong upward air currents. These clouds are known as thunderheads if they are seen during a storm. Cumulonimbus clouds can form singly, in groups, or along squall lines associated with cold fronts. Lightning and other harmful severe weather, such as tornadoes and hailstones, can be produced by these clouds. Cumulonimbus clouds emerge from overdeveloped cumulus congestus clouds and can form into a supercell. Cb stands for Cumulonimbus.

Cumulus

Cumulus are clouds with flat bases that are sometimes described as "puffy," "cotton-like," or "fluffy." Cumulo-, which means heap or pile in Latin, is the origin of their name. Unless they are the more vertical cumulus congestus variety, cumulus clouds are low-level clouds that are normally less than 2,000 m (6,600 ft) in altitude. Cumulus clouds can appear singly, in groups, or in lines (Movva et al., 2012).

When impacted by weather elements such as instability, moisture, and temperature gradient, cumulus clouds are frequently progenitors to other forms of clouds, such as cumulonimbus. Cumulus clouds typically produce little or no precipitation, but they can develop into precipitation-bearing congests or cumulonimbus clouds if they grow large enough. Depending on the ambient temperature, cumulus clouds can be created from water vapor, supercooled water droplets, or ice crystals.

They exist in a variety of shapes and sizes, and they generally cool the planet by reflecting solar radiation. Cumulus clouds belong to the free-convective cumuliform cloud family, which also includes cumulonimbus clouds. Due to its more complicated structure, which frequently includes a cirriform or anvil top, the latter genus-type is sometimes classified independently as cumulonimbi form (Müller et al., 2019). Stratocumulus (low-étage), altocumulus (middle-étage), and cirrocumulus (high-étage) are examples of cumuliform clouds with little convection. These three genus-types are occasionally classed as stratocumuli form.

Current

A current is a consistent, predictable flow of one fluid through the body of another fluid. Fluids are materials that can flow and change shape readily. Water is the most frequent natural fluid. The volume, or amount, of water flowing in a river; the river's steepness, or stream gradient; and the topography of a riverbed, which includes sand bars, basins, boulders, and dams, all influence river currents.

Cycle

A series of events that are regularly repeated in the same order.

Cyclone

A cyclone is a large-scale air mass that rotates counterclockwise in the Northern Hemisphere and clockwise in the Southern Hemisphere as seen from above, rotating around a strong center of low atmospheric pressure (opposite to an anticyclone). Cyclones are defined by inward-spiraling winds that revolve around a low-pressure zone (Metz, 2008). Polar vortex and extratropical cyclones of the biggest scale are the largest low-pressure systems (the synoptic scale). Tropical and subtropical cyclones, as well as warm-core cyclones, are included on the synoptic scale. Smaller mesoscale features include mesocyclones, tornadoes, and dust devils. During the summer months in the Northern Hemisphere, upper-level cyclones can form without the presence of a surface low, pinching off from the base of the tropical upper tropospheric trough. Cyclones have also been observed on planets beyond our solar system, including Mars, Jupiter, and Neptune. Cyclogenesis is the formation and intensification of cyclones.

D

Debris

Debris is the name given to solid waste materials produced at construction sites, such as wood, drywall, and metal. Debris can be used as mulch to slow down the erosion of soil and to make the ground harder for weeds to grow. Mulch can also prevent drought by reducing evaporation of water in the soil. In other circumstances, debris can be defined as what is left over when a person, animal or object has been destroyed by natural disasters or other catastrophes. Hurricanes. Tornadoes. Earthquakes. Pollution. These are but a few of the causes of natural disasters that cause serious damage to cities, towns, and homes on a global scale (Li and Li, 2016). Scientists have discovered that these disasters leave behind harmful debris as well, in the form of meteorites, asteroids, and even extraterrestrial rocks brought to Earth through space travel. In order to protect the Earth from debris issues caused by these disasters, NASA scientists have introduced satellites into space that track debris, including meteor dust in Earth's own atmosphere (Figure 23).

Figure 23. Debris from a fallen building.

Source: Picture by DNA India.

Decay

In Earth science, decay is defined as: The action of decaying. Rotting or change to a stinky condition. It is something that happens to everything in nature. It is defined as the gradual wearing-away of something over time. Anything that is alive or once alive, whether an animal, a bacterium, or a tree, will eventually die and decompose into its original non-living components (dirt). The rate of decay increases with temperature. In other instances, decay can be defined as the gradual process of breaking down rock or soil into smaller components such as gravel, soil, or sediment (Ledley et al., 2011). This process results in a small number of new components including minerals that form during chemical weathering. It can also be defined as the erosion of rocks and mountains, caused by a variety of factors including water and wind. This process is constant and continuous, working on all exposed surfaces over time (Figure 24).

Figure 24. Decay of forest wood.

Source: Picture by the conversation.

Decompose and Decomposition

Decompose means to break down into simpler, separate parts. That is what happens to all living things that die. Decomposers are living things that break down other dead organisms. Some examples of decomposer are fungi, salamanders, and earthworms. Decomposition is the process of breaking down dead organic materials. It usually consists of three steps: initial decay, where large molecules are broken down into small ones; continued degradation, in which smaller molecules are converted into simple compounds and gases; and final decay, which involves the breakdown of gases into thin air. The breaking down of such organic materials bring about simple substances such as humus. Decomposers are what make this possible: bacteria, fungi, and other organisms (Lewis, 2008). This process transforms dead plants and animals into nutrients that animals can use as food. Microbes and other small invertebrates are decomposers, as well as mushrooms, which are neither fungi nor animals. Decomposition occurs naturally and is the final stage in the deterioration of organic matter. The decomposition process can happen in different ways. For example, it can happen when the dead matter rots, or when it is consumed by bacteria. One way of looking at decomposition is by picturing a log rotting away. One can see the white, soft inside of the log decaying and turning into black mold as it is broken down by bacteria, molecules are oxidized.

Degree

The word degree is a noun that describes a unit of measurement in a variety of fields, including Earth science and weather. What is a degree? A degree is one part of a measuring system, including the metric system, or the Imperial measurement system. The Imperial system defines the degree as a unit of measurement which is defined as 1/360th of a full circle. Degrees are used in the field of Earth science, for example in measuring the temperature on the Fahrenheit scale (Košler, 2007). It can be defined as a unit of measurement that indicates the angle between a reference line and the vertical or horizontal position of its location. Earth scientists use scales such as the Celsius, Kelvin, or Fahrenheit to define degrees (Figure 25).

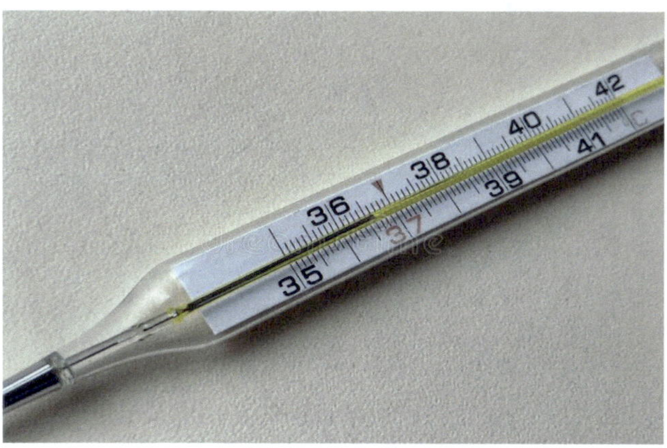

Figure 25. A thermometer measured in degree Celsius.

Source: Picture by Dreamstime.com.

Delta

Delta is the innermost of the Earth's four rocky small world bodies. It was discovered in the 19th century and named "delta," which is Greek for "fourth." Delta is a collection of connecting valve bodies and accessories designed to provide comprehensive water solutions through a single, patented interface. It is based on Earth Science, the study of the earth's surface and beliefs about the worlds beneath our feet (Kleinhans et al., 2010). Delta provides exceptional strength and performance for today's ever-evolving water challenges. Delta in terms of Earth science is a program that helps you study plate tectonics and earthquakes, understand the Earth's changing surface features and how they relate to climate change.

Dense and Density

Dense is something that has a lot of volume in a small space. Dense materials have a high weight, or density per unit of volume. Taking for instance dense soil which is good for growing trees and plants, but it will not work for your garden.

Figure 26. Jar filled with rocks is denser than one filled with ice cubes.

Source: Picture by iStock.

Being dense means being heavy because of how tightly something is packed. For example, a toy chest filled with hard-packed toys has a high density while an empty chest has low density. Rocks are dense because there are many molecules packed into them and heavy. Lighter things like cotton candy have few molecules packed in and so it is not very dense. Density is the measure of how close together the particles that make up a material are. Density depends on the material (Klippel et al., 2019). Some materials, like water or mercury, are very dense — they have a lot of particles packed into each cubic centimeters at room temperature. Dense materials are those that have a high mass per unit of volume. Density is expressed in g/cm^3 or kilograms per cubic meter. Solids can be divided into two groups, though: Solid and dense materials (those with a large density) and less dense materials. Density can be explained in a simple experiment. Fill two glass jars or glasses with water, one jar with ice cubes and the other jar with a rock. Both jars of water will have the same volume – but which has more mass? The heavy glass full of rocks obviously contains more mass than the other jar which contains water (Figure 26).

Deposit

Deposit is a natural resource. Deposits are materials left over from minerals or organic remains of the earth's crust after sedimentation process. One can ideally claim deposit as any material left on the earth´s surface.

Figure 27. Volcanic deposits at Okmok.

Source: Picture by U.S. Geological Survey.

The most common of them are: sediments, sedimentary rocks, volcanic rocks, and crystalline rocks. Deposit is the term that Earth scientists use to describe layers of material that are deposited over time, sometimes by melting and recrystallizing in a distinct layer, sometimes by covering and preserving something existing below (Hundman and Mattmann, 2017). The layers can be sediment, lava flows, ash beds, or glacial deposits (earth scientists use the term glaciogenic deposit) – whatever they are they get classified as a deposit. It can be described as a layer of sedimentary rock material that was laid down at the bottom of a body of water through the action of gravity, and this layer is used to protect some organic remains (Figure 27).

Deposition

Deposition is the name of a process by which sediments (or sedimentary rock) are formed. This process happens when something remains on Earth surface for long time and became a solid matter. Deposition on Earth happens when something is dropped onto the land or another object. This can be either a liquid or a solid that forms an outer layer around an object while it is being carried by water, wind, or something else moving around. Deposition is when a river caries sediment, or rock and dirt, down the river. It is a major source of rock and dirt that forms soil (Guo et al., 2016). Deposition is caused by natural processes, such as melting glaciers, erosion by wind or water, gravity pushing things down the hill, or even earthquakes that dump rocks into the river. Deposition can be thought of as mineral accumulation, because it involves the addition of material to land. Deposition is the opposite of erosion, which involves the removal of material from a landform.

Depression

Depression is a bleak and desolate place. It is an Earth science term that describes an environment in which the land appears to be a flat floor of grayish-brown rock and sand. This type of environment is very dry, and there is not much life present in it. A depression is a low point in the Earth's surface, such as a shallow ocean. The word depression more often refers to land, so in farming it could be used to reference any dips within the earth's surface. The severity of a depression can ranges from a hole that just reaches the floor, to needing a shovel to dig out of one. In their most severe cases depressions can have tremendous depth and capacity in which several people or animals can lose their lives due to being buried alive) (Goldstein et al., 2017). Depression is a low area, or basin, in an ocean or large body of water that typically forms around areas where warm ocean currents move against the outer margin of a continental plate. This can be caused by a volcanic eruption, high heat flow from the subducting plate above it, or the movement of cold currents (Figure 28).

Figure 28. Depression on the Earth surface.

Source: Picture by CNN.

Desert and Desertification

Desert is a hot, sandy, or rocky place that gets very little rain. At first you might think that there is no water in the desert, which would be a bad thing for us. However, that is not true! There are many animals and plants living in the desert that have to adapt to the dry conditions. Come on an adventure with me to find out more about this amazing place on Earth. Deserts cover approximately one-fifth of Earth's land surface and they cover huge areas, but they are also isolated ecosystems (Escorihuela, 2018). In a desert, there is very little precipitation and less water evaporates than in any other type of biome. There are three types of deserts: hot deserts, cold arid regions, and other types of arid areas. The Gobi Desert in Asia and the Sahara in Africa are the largest deserts on Earth. The Great Basin Desert in North America and the Sonoran Desert in Mexico share only about 20% of the territory. Weather in the desert can be extreme. A thin layer of green appears only occasionally when an area is watered. The majority of the water evaporates quickly. There are a few plants that can live in this region, but they must be able to survive on little water for long periods of time. Earth is slowly being stripped of its natural resources, whether through overexploitation of forests and their wildlife, or the razing of grasslands, and pollution of air, water, and soil (Duffy, 2014). Desertification means that periods of drought can rip through areas normally lush with vegetation for weeks or months at a time. Ultimately, deserts are expanding rapidly throughout the arid regions of earth. Desertification is the process by which certain ecosystems that are desert like become desert in fact. It can be triggered by climatic changes or human activities such as overgrazing. Desertification is a major global environmental problem and imposes huge costs for affected countries (Figure 29).

Figure 29. Desert.

Source: Picture by Britannica.

Devonian

The Devonian Period was named after a place in France and began around 410 million years ago. Volcanic islands in the shallow sea contained large amounts of limestone, which were deposited during this time. The environment was tropical and marine, with broad continental shelves, lagoons, and reefs. The fauna of the period included primitive fish, corals, sea lilies and insects (Clowes, 2010). This period propagated the transition from pre-Cambrian plants to the first early land plants in the form of vascular plants, including Archaeopteris and Cooksonic. Like the Mesozoic Era, the Devonian was a time of major dinosaur evolution, and it was also when fish developed limbs, evolving into amphibians.

Dew

Dew is a natural phenomenon that is important to the everyday lives of many people and animals. It is made of tiny drops of water-called condensation-that form on the surfaces of things when they are cooled by the air or ground. In the morning, its' on the grass, on the leaves, on your picnic blanket and even in your hair (Clary and Wandersee, 2011). You cannot see it but you know it is there by its' name. Basically, dew is a natural state of water that forms on cool surfaces. It is formed from atmospheric moisture, so small amounts of dew may be present even in the driest deserts. Since there must be no visible evidence of rain or clouds, dew usually appears on clear nights and early mornings. When the temperature falls to around 32 to 40°F, the air can no longer hold the required volume of water vapor, and some water begins to condense. This process continues until the air can only hold so much water, or equilibrium is reached between air and water density (Figure 30).

Figure 30. Dew on the grass every morning.

Source: picture by agodman.com.

Dew Point

Dew point, in Earth science is the saturated vapor pressure of water vapor at the same temperature, or dew-pressure, at which the water vapor would be present as dew on surfaces of various materials when the ambient temperature is below freezing. It is also defined for supercooled water in a state of ice crystals within a cloud. The dew point corresponds to the temperature at which this crystal formation begins to occur when a parcel of air is cooled to saturation with liquid water. The dew point is a measurement of how much moisture (water vapor) is in the air. It can be measured using different scales, but it is always given on the same atmospheric pressure scale (1 atmosphere). The higher the dew point, the more water vapor there is in the air and the closer you are to saturation (Cossu et al., 2010). This calculator displays values in both degrees Celsius and degrees Fahrenheit. This calculator gives you a good idea of humidity or water vapor content in the atmosphere based on the temperature and dew point (Figure 31).

Figure 31. Dew point.

Source: Picture by Britannica.

Diameter

The Earth's diameter is the distance across the middle. The Earth's diameter is 6,371.0 kilometers (3,947.7 miles). The circumference of the planet at the equator is 40,075 kilometers (24,901 miles).

Diffuse

Diffuse is an Earth science term meaning to spread out or break up into smaller parts. It refers to a ray of light, heat or sound that spread out in many directions. Diffuse is used in Earth science to talk about the scattering of light. Light shining toward a surface that is diffuse reflects off the surface in all directions—even those that are not visible (Cerase, 2019). Objects that have both reflective and non-reflective surfaces (like the outer planets of our solar system) are said to be largely "diffuse"—as opposed to planets (like Earth) which contain mostly reflective surfaces and therefore appear bright from a distance.

Dinosaur

Dinosaurs are prehistoric reptiles that roamed Earth millions of years ago. They were the ruling creatures on the planet until a giant meteor hit Earth, causing catastrophic climate changes. Dinosaurs eventually died out but some have been found all over the world in a fossilized state. There are 49 different species of dinosaurs, ranging in size from the three-foot long Compsognathus to the 150-ton Brachiosaurus. Dinosaurs ruled the Earth more than 150 million years ago (Britner, 2008). These prehistoric creatures roamed the Earth millions of years ago, yet their remains are still being discovered by scientists. Their remains have been found all around the world, even though they were always huge. Dinosaurs were Earth's dominant animals, and many large groups flourished during the Mesozoic era. The rise and fall of dinosaurs, and most other creatures on Earth, is constantly being revised as new data becomes available (Figure 32).

Figure 32. A dinosaur.

Source: picture by NPR.

Dioxide

The Earth's atmosphere has a mix of gases. The main gases, Orin the Earth's atmosphere are inert gases, air, and water vapor, carbon dioxide (CO_2), nitrogen, and oxygen. Dioxide is elemental oxygen. It has a greenish color and a pungent odor. It makes up most of the Earth's atmosphere, and can be obtained in sufficient quantities from air to support life (Butler, 2016). Dioxides are compounds that contain oxygen atoms bonded to a central carbon atom with two additional oxygen atoms. Dioxides may be solids, liquids, or gases and are commonly found in air. Dioxides can cause breathing problems and other health problems if you are exposed to them for a long time. Breathing smaller amounts of dioxides also can irritate your nose, throat, and lungs.

Displace

Displace is a no-mess, clean-cut way to move earth. When a volcano erupts, it can move the Earth several feet. Displacing Earth is a process in which water is poured into a selected area and allowed time to filter down below the surface. Afterward, the water can be pumped back out. This process can help remove contaminates from the ground, such as dangerous heavy metals, pesticides, and hazardous chemicals that have been improperly disposed of in the soil or landfill. When a rock is slowly pushed into the crust under high pressure, eventually it will be pushed all the way through, from surface to mantle (Bohle, 2015). This is called subduction. Imagine a rock sitting on top of a great horn, or overhang. As one tectonic plate slowly moves over another, it pushes the rock up, but when it falls back down again, it plummets into the mantle below, displacing some material there and causing an earthquake near where the plates are moving. This is the displacement of the Earth (Figure 33).

Figure 33. Rock displacement.

Source: Picture by Science Direct.

Dissolve

Dissolve is a process in Earth science that describes the movement of ions from an area of higher concentration to one where it can combine because of lower concentration. This process occurs most frequently in nature with minerals. Over thousands and millions of years, rock materials dissolve after being exposed to high temperatures. It can be described as a process of decompose or disintegration of a substance into its elemental form. The term dissolve is also used to define a type of cloud formation, namely the: Cirrus spissatus cloud formation (Bohle, 2015). Dissolve may be used to refer to the process of dissolution, in Earth science, by which rocks or minerals on, or near the surface of the Earth are slowly broken down and pass into solution. The process of dissolution, through which minerals and rock break down chemically into simple components, occurs in many environments, including groundwater, soils, and oceans.

Distributary

The headwater of a stream. A river that flows into another, but is not the main source of water for the other. A distributary is a secondary channel of a river in its flood plain, and typically both are incised into the ground during a flood, but if the water level falls enough, a distributary may become filled with sediment. It can also be described as a natural stream of water (usually freshwater) that originates in hills and mountains and flows into rivers, lakes, or the sea (Zhuang et al., 2019). A distributary is a river branch that flows away from the main channel into an estuary. Distributaries are formed in places where a river is not strong enough to break through vertical banks of sediment. They can be many kilometers long (Figure 34).

Figure 34. Distributary channels in a river.

Source: Picture by Geology Pics.

Distribution

Distribution is the general term used to describe the pattern of where organisms are located on Earth. Scientists involved in biogeography use this term to refer to how species are spread out across the world. It not only how the species are spread around the Earth but also the spread of elements such as water, air, and land. We tend to think of Earth as a big, solid rock where things just sit still. But this is not the case at all: Earth's crust and mantle are continually shifting, and the planet experiences substantial internal energy and mass redistribution.

Divergence

Divergence, in relation to Earth science, refers to the lateral movement or spreading apart of a part of the surface of the earth. Divergence occurs at a divergent boundary, where two plates move away from each other and is most often associated with mid-ocean ridges (MORBs), which are formed as new oceanic crust is created. In other instances, Divergence is described as a measure of how far from a horizontal line the tie-lines of an aerogravity map are displaced. The closer to zero the divergence, the more accurate the map. It is also the rate of movement along a fault plane. The faster the divergence, the faster the fault is moving such as the Groundwater flow (Figure 35).

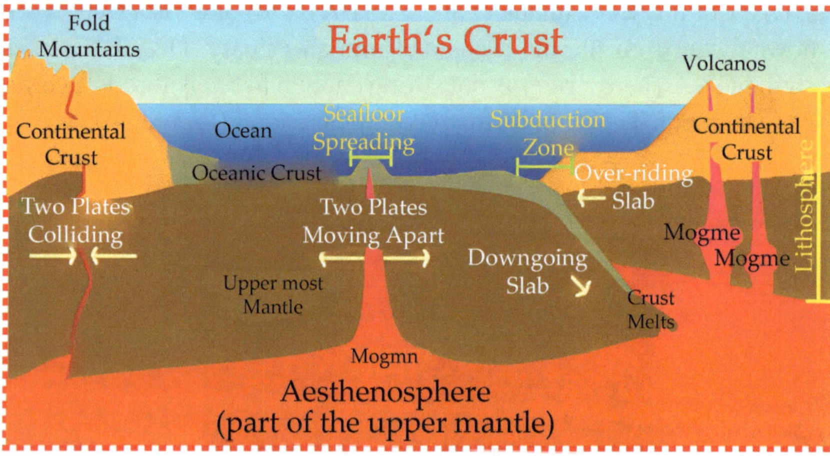

Figure 35. Spreading apart of part of Earth surface described as divergence.

Source: Image by Enchanted Learning.

Divergent

It is used in describing the differences of the earth's elements and occurrences. It can be able to describe the similarities that exists with cases in which the elements are different. The earth's circulation system is one of the most important and fundamental systems in nature. As we observe earth's atmosphere, ocean, and crust through processes such as changes in air temperature, soundings of sea waters, and earthquake analysis, we find that there are many similarities among the three systems (Yu and Gong, 2012). The cycle of water recycling between oceans and atmosphere is one example. The movement of seismic waves through the earth's interior also resembles the movement of sound waves through air.

Diversion

The distance of the shortest line between two points on a sphere, as measured along the surface. Diversion is slightly shorter than distance, but it is only a little bit. You can use diversion to say that something is not true or that someone is not right.

Doldrums

It is a state of inactivity or stagnation. In meteorology and oceanography, it is the name given to a belt, about 30° in width, within which the trade winds blow gently and without vigor. The Doldrums are the calm, warm, and sunny region of the Atlantic Ocean between the subtropical high-pressure area known as the Bermuda High and the Azores high pressure area. A doldrum is a zone of light winds near the equatorial region. The word doldrum comes from a Dutch phrase "bellen des strooms" which means "mild periods," and refers to areas where sailing ships would become trapped when sailing around the tip of South America in the 16th and 17th centuries (Witze, 2013). When sailing out of European ports particularly from Holland, England, France, and the Iberian Peninsula ships would head southwest to pick up the strong Southeast Trades wind for a direct trip across the Atlantic. During the Northern Hemisphere winter months between June and September, reaching west across tropical.

Doppler and Doppler Effect

Doppler is described as a change in the frequency of a sound or light caused by movement of the source, the listener, or both; it is used in weather forecasting and to detect movement in space. The world's first portable weather radar. The doppler is a high-performance phased array weather radar offering pinpoint radar detail, and total versatility. Using Doppler, you can detect rain, reflectivity, wind direction, wind speed, distance to object, temperature, barometric pressure and more. The doppler effect is a bit of science that shifts the frequency of waves as they travel. It is what explains how bats and dogs can use sound to find their way around. You have likely heard that long distance electromagnetic radiation always moves at the speed of light, no matter the source or distance (Wilkinson et al., 2020). But what if the source is moving? Similar to how a train whistle appears to change pitch as it approaches and then recedes from your position, an electromagnetic wave from a moving object shifts toward a higher frequency. This effect is referred to as the doppler shift, and was predicted by an Austrian physicist named Christian doppler in 1842 (Figure 36).

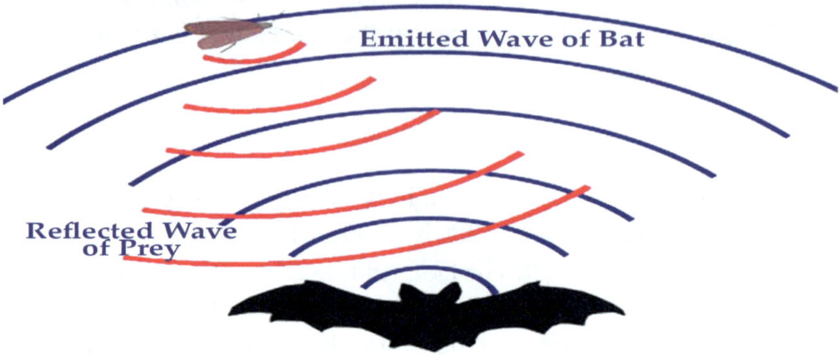

Figure 36. Bats use the doppler effect.

Source: Image by CK-12.

E

Earth

The word Earth is a name given to the planet we live on and refers to its size, shape, and location among the other planets of our solar system. Earth is the third planet from the Sun, and it is more spherical in shape than any other planet. A big benefit of living on Earth is that it has liquid water. Earth is one of the four fundamental states of matter (the other three are air, water, and fire). Science related to Earth includes studies such as oceanography and meteorology. The word Earth comes from the Anglo-Saxon word Erda, which means ground (Wallace-Wells, 2019). Earth is the third planet from the sun. Given the fact that it is the only planet with liquid water on its surface it has made it a suitable environment for humans to settle on it. Earth's atmosphere is mostly made of nitrogen and oxygen, but there are also small amounts of carbon dioxide (CO_2), methane, water vapor and other gases. The physical composition of the Earth can be broken down into three major layers: lithosphere, hydrosphere, and atmosphere. The lithosphere is the layer of Earth that includes the crust, the upper part of the mantle, and part of the lowermost mantle. The crust and the uppermost mantle make up the lithosphere. The majority of this article focuses on the properties of this layer. The lithosphere is divided into several tectonic plates which are the sections on which rest oceanic basins and continental masses (Figure 37).

Figure 37. The planet Earth.

Source: Picture by The World Economic Forum.

Earthquake

Earthquake is the shaking of the surface of the Earth caused by the sudden release of energy in the earth's crust. An earthquake is basically a tremor producing shakes which cause disaster to human beings. This happens when two blocks of the Earth rub against each other. It can be described as a sudden, but usually brief, release of energy in the Earth's crust that creates seismic waves. These energy waves travel through the Earth and cause the ground to shake, sometimes violently. The point on the Earth where an earthquake begins is called its focus or hypocenter (Toth et al., 2014). They tend to happen when plates under the earth's surface pull apart or move suddenly, causing a giant release of energy and violent shaking. Most of the time, this energy is released deep below the surface; however, sometimes stress causes a piece of the earth's crust to break off. They are usually small, so you do not notice them. Los Angeles has a lot of earthquakes because it is near a fault. Earthquakes are powerful enough to make buildings sway and rumble, but rarely do they cause total destruction. Even in the worst-case scenario, which is an earthquake that hits at night when many people are asleep, most earthquakes do not have a devastating impact on life outside of the immediate area (Figure 38).

Figure 38. Earthquake occurrences.

Source: Picture by EquityPandit.

Earthy

Earthy is a collective term that refers to all things related to the soil, including its mineral and physical components that influence plant growth and health. It is a reddish-brown natural pigment.

Echo Sounder

An echo sounder is a type of sonar that was developed in the late 19th century. It uses radio waves to measure the depth of water and other solid and liquid surfaces. An echo sounder is made up of a transmitter, which emits a pulse of high frequency radio waves at regular intervals, and a receiver, which detects the reflected echo when it bounces off of a solid or liquid surface. Furthermore, it measures the speed and distance of vessels, or to make special sonar measurements needed by underwater researchers, such as studying vegetation or fish species. It consists of an electronic device, a transducer, and usually a very long cable. The transducer has a transmitter on the surface and a receiver under water connected by the cable. Time passes more slowly in water than in air, so for example if the pulse travels for 10 ms (1/10s) to travel there and back to the receiver, the distance from the device to the bottom will be around 100 m (Thessen, 2016). The calculation is based on the speed of sound in water which is about 1,500 ms/km. The device has also been used by fishing experts in locating fish by sending out sound waves and recording the reflected sound waves by means of hydrophones which are devices sensitive to pressure changes.

Eclipse

An eclipse is an astronomical event that occurs when an astronomical object is temporarily obscured, either by passing into the shadow of another body or by having another body pass between it and the viewer. On the Earth surface it occurs when the Moon passes between Earth and Sun, blocking our view of the solar disk. This can happen only during a new moon, when the Moon's orbital plane about Earth lies in a line with the Sun as seen from Earth. Earth casts two shadows into space, one north to south and one from east to west. The eclipse is just outside the umbra or inner shadow because at that location skies are always Darkest. The Moon's silhouette eclipse is easily visible during a total solar (Telling et al., 2017). On Earth, this is a rare occurrence. Even though it can be seen from anywhere on Earth where the Sun is up, it is possible to see an eclipse only in places where the Moon's shadow falls. The shadow of the Moon can be seen as a cone-shaped region with its apex directly opposite to the Moon (Figure 39).

Figure 39. An eclipse.

Source: Picture by Allure.

Ecliptic

Although not visible to the naked eye, the ecliptic is distinct in the night sky. Eclipses, comets, and meteors appear along a band of the celestial sphere 60° in front of the sun. This band is called the ecliptic since eclipses occur when Earth passes through it. The ecliptic forms an angle of 23 1/2° with respect to the celestial equator. It crosses the planets from south to north as they orbit the Sun (Showstack, 2008). The point on Earth directly beneath this path of intersection is known as the subsolar point. It is roughly in the same direction as the zodiac and corresponds to a great circle passing through both poles. The ecliptic is not identical with the celestial equator, which is inclined by 23.4° from the ecliptic plane. Like the Sun, ecliptic is present only in the tropical zodiac. However, it is not a fixed line. It slowly rotates across the 12 constellations of the tropical zodiac at the rate of about 1° every 72 years (Figure 40).

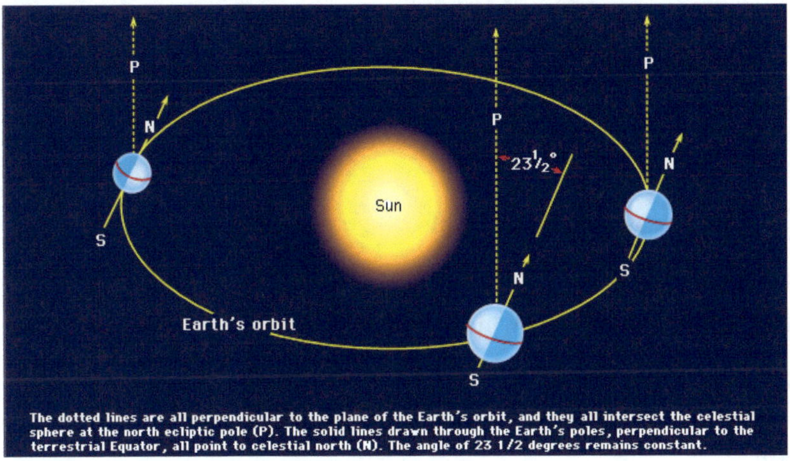

Figure 40. The ecliptic phenomenon.

Source: Image by Britannica.

El Nino

It is a phenomenon occurring on the Earth. It occurs in all stages of the development of the atmosphere, and also has an effect on human activity. The El Niño effect is a phenomenon that occurs on the Earth's oceanic and atmospheric conditions in the Pacific Ocean and its tropical atmosphere. The effects of El Niño are global. It is associated with periodic changes in weather patterns throughout the world, such as wind, rainfall, and temperature over substantial areas of the global oceans, including the Pacific and Indian Oceans. It is characterized by a strong fluctuation of air. It involves oceanic warming, which results in a change in the direction of the trade winds that circulate tropical waters (Schmitz, 2011). This shift in trade winds changes weather patterns to those normally experienced during El Niño events. Its effects vary regionally, with wind and rain along the American coasts, droughts in Australia, Indonesia, and India, dry weather in Central America, heavy rains in South America's Amazon Basin, and warmer than normal water temperatures as far as Africa (Figure 41).

Figure 41. The occurrence of an El Nino.

Source: Picture by fao.org.

Elastic

It is the ability of a material to regain its size and shape after pressure is removed, similar to the stretching and bouncing behavior of a rubber band or spring. This is what keeps the Earth's surface from falling into the center of the Earth. The planet's surface is broken into several large pieces called tectonic plates. When two of these plates slide against each other, pressure builds until the rock breaks and earthquakes occur (Schmitz, 2012). Researchers say that an elastic layer exists about 100 kilometers (62 miles) beneath the Earth's surface. At this depth, the enormous pressures and temperatures are thought to cause water to turn from a liquid into a kind of 'super-hardened' solid.

Electrical

The electrical Earth, also known as communis or Tellus, is one of the four classical elements in ancient Greek philosophy and science. It was commonly associated with, and named after, the planet Earth, and is thus also referred to as terra (Latin for Earth) in world languages other than English. The word "electrical" comes from the Greek word for amber (Rao, 2012). The electrical conductivity of the Earth is the electrical conductance between the Earth's interior and outer conductor. It is a quantitative representation of the electric currents flowing from the Earth's interior to its surface, and likewise of currents flowing in the opposite direction.

Electromagnetic

Research in the EM field includes investigations of the Earth's magnetosphere, measurements of environmental electromagnetic fields and studies of the impact of atmospheric electricity on radio wave propagation, lightning, and airglow. Earth is surrounded by magnetic fields produced by the flow of electric currents in the liquid outer core, and in the solid iron inner core (Rivet and Kastens, 2012). The study of these magnetic fields is known as geomagnetism or aerology, and the magnetic measurements are known as "geomagnetic measurements." The North geomagnetic pole, South geomagnetic pole, and data about other locations on Earth such as physical water body formation are recorded for use in global navigation systems such as GPS. Electromagnetic force (EMF) is a physical phenomenon which attracts electrically charged particles and induces electric currents in nearby conductors. The EMF is one of the four fundamental forces of nature. The others are the strong force, the weak force, and gravitation.

Electromagnetic Radiation

A weak form of energy traveling in space that is propagated by changes in the electric and magnetic field. Electromagnetic radiation has energy ranging from gamma rays to radio waves. Electromagnetic radiation is the transfer of energy through space, usually by oscillating electromagnetic fields. The electromagnetic spectrum covers a wide range of frequencies. The ionizing form of electromagnetic radiation is considered to be particularly dangerous because it can lead to what are known as ionizing radiation and radio wave sickness (Pyle et al., 2018). It is used to describe different types of electromagnetic waves that exist on the electromagnetic spectrum. Radio waves, microwaves, infrared light, visible light, ultraviolet (UV) light, X-rays, and gamma rays are all types of electromagnetic radiation. Lightning is one form of electromagnetic radiation. A flash of lightning is viewed as a huge spark that transfers enormous electrical energy in a split second over a wide area. The region within the atmosphere where this discharge occurs is called the "plasma channel;" it can be several meters wide, reaching temperatures exceeding 30,000 K (27,778°C) and producing currents that typically peak at around 100 million amperes (100 megamperes) (Figure 42).

Figure 42. Lightning, an example of electromagnetic radiation.

Source: Picture by National Geographic.

Electromagnetic Spectrum

The electromagnetic spectrum is an aspect in the field of Earth science. It is a distribution or range of electromagnetic radiation frequencies.

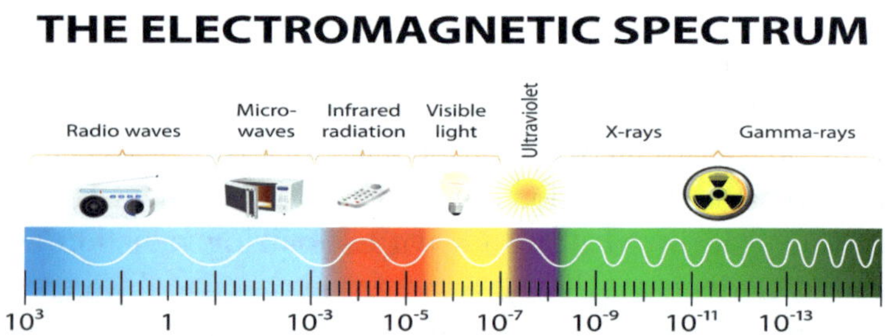

Figure 43. Electromagnetic spectrum.

Source: Picture by VectorStock.

Usually, the electromagnetic spectrum refers to radiation ranging from the longest wavelength gamma rays, through X-rays, UV light, visible light, infrared, and the entire range of wavelengths to the shortest wavelength radio waves. It is considered to be a subset of the entirety of electromagnetic radiation (Press, 2008). Electromagnetic radiation is present in both natural and man-made sources, it is contained in all photons, and plays a critical role in influencing the transfer of energy in all parts of the natural environment. The "electromagnetic" part of the name refers to EMF, which is itself generated by moving charges: electrons in motion generate a magnetic field, while atomic and subatomic particles moving through space generate an electric field (Figure 43).

Electromagnetic Wave

It is a wave of electromagnetic radiation that oscillates or travels in a path perpendicular to the direction of propagation. The electromagnetic wave is mediated by the electromagnetic field, which travels through the medium of empty space. Electromagnetic waves are a form of energy which travel through space and can be detected by the human eye. They are not found in every environment, but occur in relationship to the elements they pass through, including organic materials and soil (Plotnick et al., 2014). It consists of oscillating electric and magnetic fields that travel through space at the speed of light. Electromagnetic waves from the Sun are responsible for creating Earth's weather and climate. These waves transport energy from one point to another, with some types traveling at the speed of light (Figure 44).

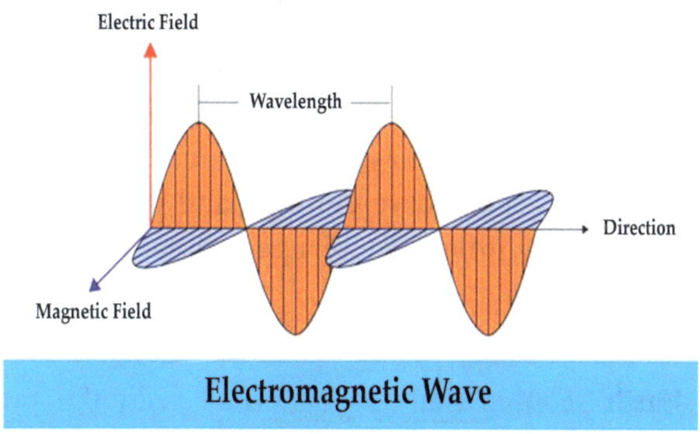

Figure 44. Electromagnetic wave.

Source: Image by Shutterstock.

Electron

Electron is an aspect of Earth science and a charged particle that orbits an atom's nucleus. This electron is approximately 1,836 times lighter than the proton, and it has one negative charge. Electrons not only orbit around the nucleus, but they also rotate clockwise around the nucleus if viewed in a positive-to-negative direction (N to S) left to right. It orbits at an approximate speed of 2×106 m/s nearly perpendicular to its orbital plane. The negative energy of an electron is offset by its negative mass, thus creating an overall neutral valence for the atom or molecule. Electrons are part of the air around us. They are tiny particles with a negative charge. They circle the nucleus of atoms and fill space inside of atoms (Pennington et al., 2020). Electrons are found in many materials, including all metals. The number of electrons in an object determines its chemical behavior and physical properties, such as its ability to conduct electricity. It can be described as a subatomic particle found orbiting the nucleus at the heart of every atom. It is similar to a tiny magnetized loop of electrical current that orbits the nucleus, and it has a negative charge.

Electronic

It is an aspect that encompasses the use of electricity, electromagnetism, and electromagnetic fields to quantify and study matter. Electronic describes things related to electricity and the behavior of electrons. As technology advances, electronics are updated and put to use in devices functioning in many different ways. Examples of electronic devices include televisions, calculators, radios, telephones, and microwaves (Perdigão et al., 2020). It is known to be a branch of geoscience involving the study of ancient magnetic fields preserved in rocks. Geomagnetic data are routinely captured at specialized observatories and archives. Data are also gathered in the field by experienced geophysicists, amateur magnetometers and through systematic investigation of rock exposures, core samples, and studies of paleomagnetic null hypotheses for areas with no obvious geomagnetic information.

Emission

An aspect of Earth science that is concerned with the production and behavior of gases in the Earth's atmosphere such as CO_2, methane, sulfur hexafluoride, ozone, and water vapor. Earth is home to a complex network of life and non-living systems. These systems interact in ways that change the planet--and they were created by a host of sources, which release hazardous emissions. In combination with the right type of atmospheric conditions, these emissions create climate change and affect Earth's natural responses to it. Such gases are described as air pollutants. The internal combustion engines of cars, boats, and aircraft, have been linked to smog, acid rain, and other changes in the environment. CO_2 and water vapor are among the most common "greenhouse" gases. Greenhouse gases trap heat in the Earth's atmosphere. As more greenhouse gases build-up, less heat escapes into space (Figure 45).

Figure 45. Greenhouse gas emission.

Source: Picture by LiveScience.

Emission Spectrum

The emission spectrum is an aspect of Earth science and meteorology that studies the behavior of gases when heated for measurement or analysis by a spectrometer. This study is important in relation to the greenhouse effect, ozone layer depletion, and air pollution. It is a graph that describes the frequencies of light emitted by a substance that has been excited by photons. This process, in which energy is emitted in discrete units, is described as quantum emission (Parkes and Morris, 2001). These spectra can be used to identify and characterize substances with emissions lines. Road signs are also produced on film that is sensitive to radiation and inks (especially lead-based) so they can be read during the day or at night by automobiles. The emission spectrum is viewed by astronomers as a definitive diagnostic tool for determining a star's chemical makeup. This is because elements found in the solar atmosphere will either favor the appearance of certain lines, or inhibit their appearance.

Emit

The word "emit" means to give off. To emit light, for example, is to shine light. In Earth science, the word often refers to various types of materials that are evaporated from plants or animals and released into the air (like salt from seawater). Evaporation and emission also occur in volcanoes and when water freezes into ice. Emissions are the result of chemical reactions between gases which are emitted into our atmosphere. Emissions can have a large impact on Earth's temperature as some gases trap heat in our atmosphere; while others allow radiation to pass through, cooling our planet (Figure 46).

Figure 46. Carbon gases being emitted.

Source: Picture by Dreamstime.com.

Energy

Energy is a general term for the capacity a physical system has to perform work. Energy exists in many forms and can be transformed from one form to another. It is the potential of a physical system to perform work on other physical systems (Özyurt et al., 2021). Energy occurs in various forms, including various types of kinetic, potential, thermal, and electrical energy. It includes the Sun and moon, their energies that are created by the Sun, and they are affected by the moon. These two energies combine to create YARA, a universal energy which is responsible for all life on Earth.

Environment

The environment can be defined as the surroundings in which people live and the way in which we interact with the rest of our ecosystem. While it is important that we consider how we can impact our world, it is equally important to ensure that we understand the nature of our relationship with our own environments.

The environment field is just one way to study interactions between humans and their environments. There are many ways to help the environment. Driving less, recycling newspapers and food containers and not littering are a few examples.

But one of the most important things we can do is to make realistic changes in our lives, such as replacing products with a more environmentally friendly version if it costs no more money (Orion, 2019). Since its founding in 1957, the Environmental Protection Agency (EPA) has worked to protect human health and the environment. Working with state, local, and tribal authorities, businesses, and the public, EPA enforces regulations that protect air quality, drinking water quality and environmental quality overall. The agency also works with the media to track environmental issues affecting America.

Environmental

Environmental means the study of the Earth and the environment around us. Our planet is made up of six spheres: the lithosphere, the hydrosphere, the atmosphere, the biosphere, the pedosphere, and the noosphere. A study by the EPA indicates that environmental concern is greater now than ever before. From global warming and deforestation to development and pollution, there are many issues that need to be addressed.

Figure 47. The environment.

Source: image by Wikipedia.

Environmental changes can be measured in many ways, including changes to the chemistry of various elements, atmospheric conditions and climate, temperatures, and energy flow within the biosphere (Oreskes, 2015).

The environmental sciences are concerned about the relationship between human beings and their natural environment.

They study how population growth, industrialization, and consumption of resources affect ecosystems. Environmental scientists work to find ways to improve these effects or else to rectify problems that may be caused by them (Figure 47).

EON

EON is a symbol used to represent the various stages of the Earth – Precambrian, Paleozoic, Mesozoic, and Cenozoic. The Precambrian spanned from 4 to 1.6 billion years ago. The continents started to gather together. There was a lot of life in the oceans. The atmosphere consisted of methane and ammonia. The Paleozoic Era was the first era of the Phanerozoic Eon, spanning from about 542 million years to about 251 million years. Also called the "Age of Invertebrates" because over 70% of living species were invertebrates. The Mesozoic Era began 252 million years ago. That was when reptiles emerged, including dinosaurs. Planet Earth would never be the same (Oreskes, 2013). For the first time in 250 million years, a single habitat contained creatures large enough to dominate its landscape. The Cenozoic age lasted from 65.5 million years ago to 2.5 million years ago. We believe in the same natural energy force that has kept the Earth spinning around the Sun ever since its formation. That force resides within all of us and is fueled by our passion for building powerful, creative products that are made from quality materials (Figure 48).

Figure 48. Precambrian stage of the Earth.

Source: Image by PowerShow.com.

Epicenter

An earthquake occurs when rock beneath the Earth shifts and causes a rupture in the Earth's crust. A huge amount of energy then radiates out from the source of the earthquake, causing a sudden shaking of ground and buildings, usually felt for large areas around the quake. An epicenter is the point on the surface of the Earth's crust directly above the location where an earthquake originated and is located (Figure 49).

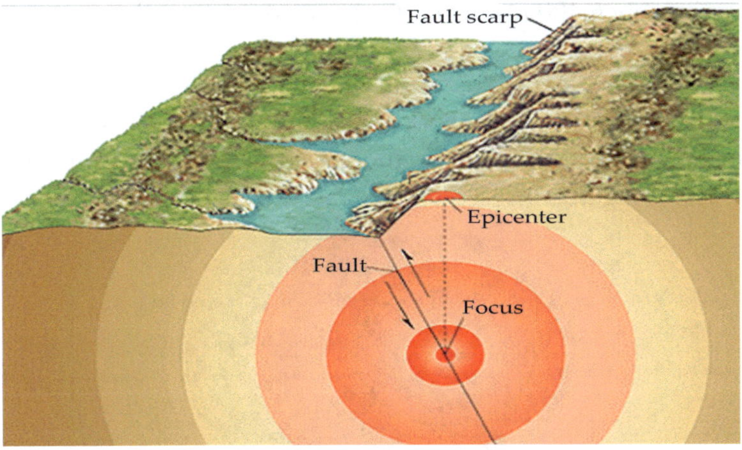

Figure 49. Location of the epicenter.

Source: Image by Instructables.

Epidemiology

Epidemiology is a branch of science that deals with the incidence, distribution, and possible control of diseases affecting large numbers of people. Epidemiology presents an expansive overview of the changes that have taken place in our world over the course of centuries.

Epoch

The epoch is the time a place exists under certain conditions and when life evolves. The epoch of a region is the duration of time under a distinctive set of environmental conditions. During an epoch, life may ascend from microbial to multicellular organisms. The speed of evolution depends on the climatic conditions of the period (Odenwald, 2020). The great evolutionary changes that occurred throughout the Precambrian took millions or tens of millions of years to produce our modern planet.

Equator

The equator is an imaginary line around the center of the Earth. It is the line that divides the Earth into two halves, in which the Sun rises in the north and sets in the south. Earth is not perfectly spherical, which makes an equator somewhat "line of sight." There is a range of latitudes where the Sun can be directly overhead, a point known as the Tropic of Cancer and Tropic of Capricorn (0° latitude). The North and South Poles are each located at 90° (or approximately 6,400 nautical miles) from these latitudes yet on opposite sides of the planet (Neeck and Volz, 2011). In other words, if you were to travel toward the North Pole from the nearest point on the equator, you would go from a point where there are no seasons (every-day temperatures are tropical), to one with very extreme seasons (every-day temperatures are freezing cold) (Figure 50).

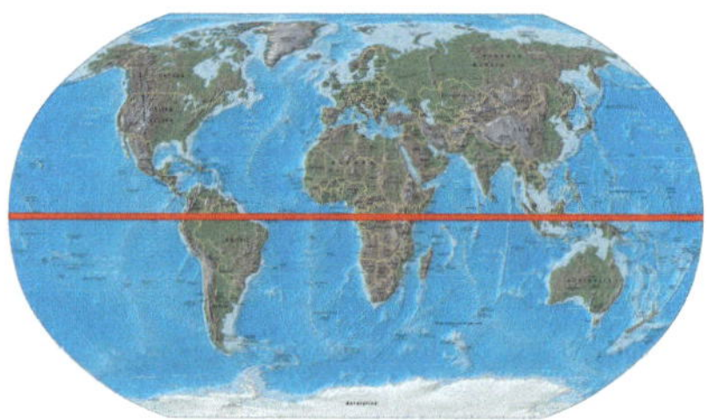

Figure 50. Equator.

Source: Image by Britannica.

Equilibrium

Equilibrium is a state in which opposing forces are balanced. The balance can be seen with the progression of energy along various pathways in the Earth system, such as water movement and winds.

Equinox

The Earth science concept of equinox describes two days a year when the Sun is vertically aligned with the Earth's equator, causing night and day to be of approximately equal length.

F

Fahrenheit

The temperature scale most commonly used throughout the world. It is based on two fixed points, the temperature at which water freezes (32°F) and the temperature at which water boils (212°F). Fahrenheit temperatures are expressed using the unit "degrees" or the letter "F." Fahrenheit was the first temperature scale used widely in the world. It was developed by a German physicist, Daniel Gabriel Fahrenheit, who lived from 1686 to 1736. In his work, he tried to determine a temperature scale that would be of the same form as the pressure-temperature scale discovered earlier in about 1650. He defined the scale's set point as 0° (the boiling point of water at sea level), and its degree increments as one-half the difference between the ice and boiling point of water (Movva et al., 2012). By comparison, the Celsius scale is based on a proposed 1601 by the Swedish astronomer Anders Celsius (1701–1744). Compared with the Celsius scale, used widely in scientific work around the world, the Fahrenheit scale is sometimes more familiar to Americans because it is commonly used in the United States; also, on the Fahrenheit scale, water freezes at 32° and boils at 212°, rather than 0 and 100° as on the Celsius scale (Figure 51).

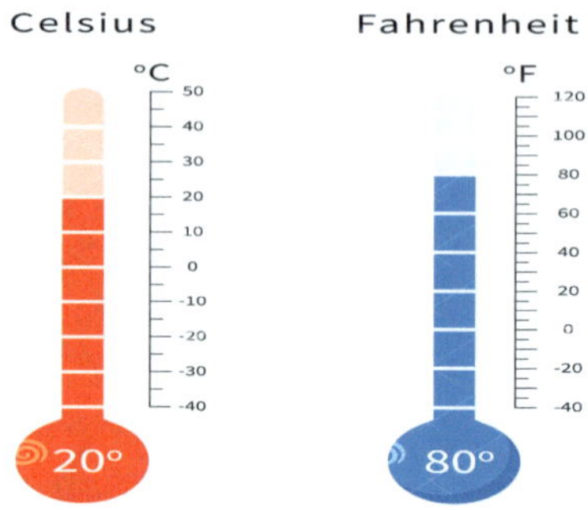

Figure 51. Celsius and Fahrenheit measures.

Source: Picture by iStock.

Faucet

Water is one of the most commonly occurring substances on our planet, as it makes up 70% of the Earth's surface. Water covers nearly three-fourths of the Earth's surface, with 97% being ocean water and only 2% freshwater. The word faucet comes from the Latin word for "to flow." The planet's tectonic plates are always shifting and causing earthquakes that can result in tsunamis. Most tsunamis happen in the Pacific Ocean and are caused by megathrust earthquakes, which means they have a magnitude of at least 9 on the Richter scale. The world's largest recorded tsunamis were from megathrust quakes that happened along subduction zones – oceanic trenches where one plate gets pushed under another and sinks into the Earth's interior (Müller et al., 2019). A faucet, also known as a tap, is a valve installed in the counter of water sources such as rivers and lakes, or attached to a wall underneath a sink. A faucet sends out a stream of water that can be controlled using a control handle. Toilets use separate small faucets to send out water. These devices allow the controlling of water flow in sinks, bathtubs, and other vessels by opening, closing, or partially obstructing various types of valves. It uses the process of sub-pressure to regulate the pressure above atmospheric pressure, which makes it possible to control and avoid the splashing fountain water with its high speed. The water will drop when pressure is lower than atmospheric pressure, and flow when pressure is higher than atmospheric pressure. Contemporary faucets may be made of plastic or metal, often with handle-operated tap mechanisms and integrated spout designs. Faucets are also used to dispense other fluids such as hydrogen peroxide or Divine's fluid from constructo-tanks.

Fault

Underground earthquake faulting is a major source of earthquakes, especially in places like California. Faults are breaks in the thin, outermost layer of the Earth, along which movement has occurred. Faults are the surface expression of a deeper process in which one section of a planet's crust slowly moves relative to another section. In other words, a fault is a geological feature that occurs along the boundary of two geological faults. The dip angle of a fault plane, which defines the orientation of the fault plane with respect to the horizontal, is a principal factor that determines how slip occurs on a fault (Metz, 2008). The dip angle and throw of a fault can be measured in a number of different ways, and this may give rise to significant variations in estimation for given fault properties. Physical feature where the Earth's solid outer layer (crust) is standing above the mantle. Commonly caused by fault movement and accompanied by seismic activity, a fault can threaten human settlements, rupture critical water lines, sabotage oil wells and pipelines, or cause other disasters. The amount and type of displacement is usually expressed by the term fault plane (or just plane). Most faults are considered active at one time or another, but maybe on either side of the "continue" block. Since faults are the most important factor in earthquake generation, understanding them is essential, especially when there is a need to protect oneself from cases of earthquakes (Figure 52).

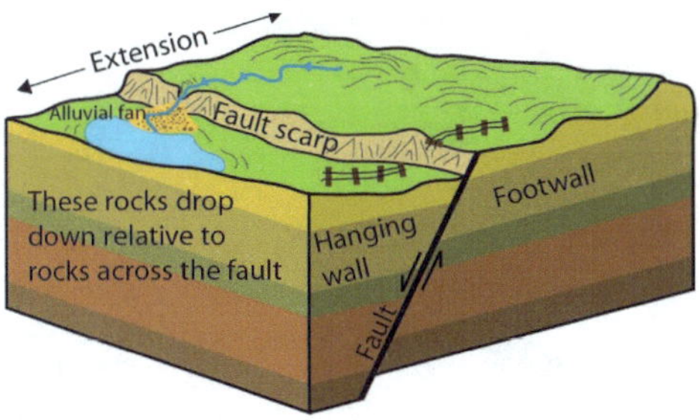

Figure 52. Faults and fractures on the surface of the Earth.

Source: Image by National Park Services.

Firn

Firn is a layer of compacted snow that has built up over time from being compressed by the weight of new snowfall. Firn is a general term for all ice that is found above sea level, snow on the ground and in glaciers. Firn comes from the German word firne which means 'old snow.' It refers to the stage of accumulation of granular snow with properties similar to those of Glacial ice: it becomes dense and hard in a process known as Glaciation. The process of new snow accumulating and aging into firn is called "firnification." It is also sometimes referred to as "old snow" or "old ice." Also known as névé, firn comes from the Latin word meaning "new," which is appropriate because it forms as winter snow gradually builds up in the winter months (Li and Li, 2016). It contrasts with older snowfields, ice wedges, and permafrost deposits. It is formed by the compaction of snow with freezing and thawing. It is denser than liquid water because of its increased cell size and solid properties. This term is also used to describe all snow and ice found in regions covered by at least 1 million square kilometers of snow and ice. A firn core is the product of this accumulation process and consists of compressible snow particles, some air pockets, and some firn—firm under layers that can be as much as 30 feet thick. The ice in a firn core is contained in pores and crevices between the firn grains and also forms distinct layers within it (Figure 53).

Figure 53. Firn.

Source: Picture by Britannica.

Fissure

Fissure is an aspect of Earth science referring to any large fracture in the surface of the Earth. This could be a crack, crevasse, fissure vent, sinkhole, canyon, or rift. The fissure, or crack, is a basic feature of the Earth's surface. It is an opening or break in the crust of the Earth, although it may not be visible on the surface. Fissures are common at convergent plate boundaries and within the interiors of plates. A crack in the ground will release heat from the depths, if it is large enough. The resulting geyser-like effect causes a thermal spring. Fissure is a geological term for a crack in the Earth's surface resulting from the displacement of two sides coupled with stress. This displacement can be caused by tectonic activity, volcanism, erosion, or simply the cooling of certain types of igneous rock (Ledley et al., 2011). The study of fissures is called fissure analysis and is an important part of structural geology and tectonics. Fissures are found in other places besides the Earth; they also exist in ice and other natural media such as sand, rock, and snow. Fissures may help to create new chasms, or they can be created when existing chasms open up further. This can happen when magma is pushed through them, but if it misses the original gap it will wind up in a new fissure elsewhere in the process known as palingenesis. Fissures may be unique features, or they can be associated with larger cracks, such as rift valleys (Figure 54).

Figure 54. An Earth fissure.

Source: Picture by AZGS University of Arizona.

Flammable

Flammable means that something is capable of being easily ignited and burned with rapid oxidation or combustion. The word "flammable" is used in chemistry to describe materials that easily support combustion. Materials capable of burning significantly faster than their supply of oxygen are the most dangerous, and the most flammable. Coal is flammable; water is not. The term flammability refers to how easy it is to ignite a substance. The word is derived from the Latin word " flamma ," meaning flame. When speaking about fire safety, we generally talk about preventing ignition of materials that might burn, but in other contexts, it is often important to make sure that a material does not ignite too easily (Lewis, 2008). For example, if you are working with chemicals, you want to make sure that none of them will catch fire or explode under normal circumstances. The vapors of most liquids are flammable, which means that they can form the visible vapors, which we call flames. Gases are also flammable, because the gas molecules can evaporate into their gaseous phase while being heated. However, some gases may be less flammable than others when they are in their gaseous phase (for example, chlorine gas is far more explosive than another highly toxic gas such as natural gas). Flammable fossil fuels are liquid fossilized materials with a high proportion of hydrocarbons that can be burned. Chronic exposure to materials or foods containing flammable gases can result in dizziness, respiratory system disorders, cardiac irregularities, and even death (Figure 55).

Figure 55. Flammable fossil fuel.

Source: Picture by The Guardian.

Flank Eruption

It is the sudden explosive uprising of molten rock, ash, and other volcanic materials onto the surface. The process creates large lava flows instead of sending hot gases and ash into the sky. One type of flank eruption can release 600 cubic kilometers of lava within 17 days.

A flank eruption, also known as a volcanic eruption, is a type of volcanic eruption, sometimes known as a dirty thunderstorm. It occurs when significantly molten lava rises through the Earth's crust, and the pressure exerted by the rising mass results in the rock above it splitting apart, often producing spectacular displays of lava fountains (Košler, 2007). Flank eruptions are short-lived compared with Shield volcano eruptions, Vulcanian eruptions or Plinian eruptions. Under certain conditions, rocks on the part of a face of a volcano located below the normal heights of eruptive activity undergo expansion due to heating and softening, which causes cracks to form between them.

The resulting stress may lead to an explosive release of built-up volcanic gases, lava, and fragmented rock. This eruption may occur along a fissure or along a portion of a volcano's cone where it meets the Earth's surface. They can also be caused by decompression through over-pressurization (blanket) lava flows, direct injection of magma into shallow crustal levels or from shallow ascending magma that encounters water. How flank eruptions occur is through the release of pressurized gas and lava, which occurs through small, preexisting vents because the volcanic edifice that is unable to contain the pent-up energy any longer.

Figure 56. Flank eruption.

Source: Picture by Wikipedia.

This leads to an explosive breakout of the accumulated lava to relieve the pressure (Kleinhans et al., 2010). When such eruptions take place in places with glacier or ice sheets, it leads to the thawing of the surrounding ice which causes it to melt or sublime. The resulting lahar can flood valleys and process huge amounts of sediment. This volume of water may be discharged from the volcano's crater, creating a pyroclastic flow possible. Although many people think of volcanoes as having an explosive summit eruption, most large volcanoes throughout the world are actually made up of numerous eruptive centers (Figure 56).

Flint

The first use of the term 'flint' in English is from 1565, although the German word 'flint' appears very slightly earlier, perhaps by no later than 1503. The Middle English forms are derived from Old Norse flint; this should not be confused with the modern meaning of flint, which only became associated in the early modern period (since about 1600) with "shining fragments of rock struck from a flintstone."

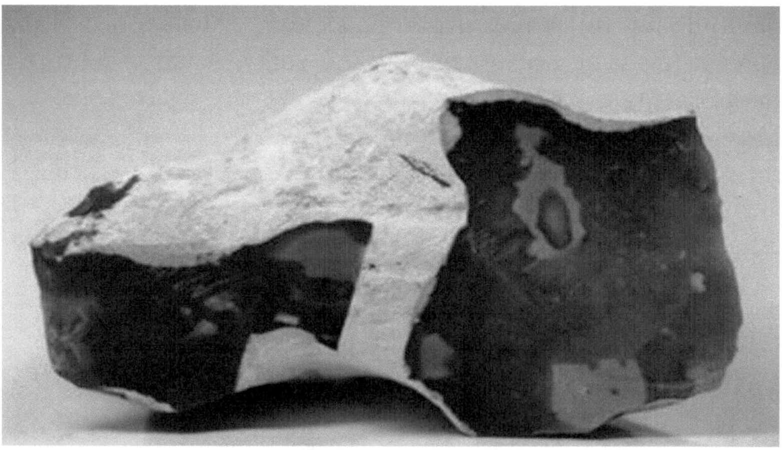

Figure 57. A flint.

Source: Picture by Geology.com.

Up until that time it simply referred to stone in any form. Flint is a hard, sedimentary rock formed from the hardened remains of seashells and the bones of aquatic animals. It has been used in tools since prehistoric times, but its modern use in gun flints for firearms starting in the 17th century reduced its use as a toolmaking material (Klippel et al., 2019). Most individuals recognize flint as a rock or piece of stone that contains a source of friction and when struck with steel generates a spark. The discovery of flint was one of the pivotal events in prehistoric archaeology. Making it the first stone age technology and with implications jealously guarded by modern archaeologists, who preserve the notion that Homo sapiens invented nothing new during their first 200,000 years on Earth. Flint occurs chiefly as nodules and masses in sedimentary rocks, such as chalk, where it is formed as a result of chemical change during fossilization, and is sometimes found in association with iron ore and coal. The term is also commonly used to refer to all forms of igneous rocks except for the extrusive and basic (silicate) rocks (Figure 57).

Flood

Flood is a noun. It means a large body of water that has overtaken surrounding land, whether by the flow of a river or by the rising of an ocean. Floods may be caused by storms, earthquakes, volcanic eruptions, ice damming or a sudden change in the level of water far upstream.

Figure 58. Floods in a town.

Source: Picture by Bloomberg.com.

Flooding may occur due to rainwater runoff from the land surface, and can be amplified if the drainage basin cannot carry all of the rainwater produced. In most cases, flooding occurs when a river overflows its banks or if water enters a land area that is not normally wet, two different but related phenomena. It can submerge land, destroy homes and property, and wash away vegetation, bacteria, and chemicals (Hundman and Mattmann, 2017). Floods are primarily caused by heavy rain or unconsolidated snow that melts. While the size and causes of some floods are very specific, others may be highly variable; they may occur only after a heavy rain, and they may coalesce with other minor floods to produce flash floods. Flood and flood plain management are a priority for local governments in areas prone to flooding. They can result from a combination of intense or long-lasting precipitation, bedrock that cannot absorb water fast enough, and topography (layers of rock) that causes water to overflow (Figure 58).

Flood Plain

The name given to areas of land that lie between the banks of the river and an adjacent higher area of land, and are filled with water when the river floods. In Earth science, a flood plain or floodplain is an area adjacent to a stream or river, bounded on either side by higher ground. The term can be used to describe a low-lying tract of land adjoining any flowing body of water, including rivers, streams, canals, and estuaries. Both the words "plateau" and "peneplain" are sometimes used in exactly the same sense as "flood plain," though the word "peneplane" tends to be more commonly used with reference to relatively flat expanses of land (Guo et al., 2016). It is normally a flat lowland adjoining a stream or river that is frequently inundated or covered with water. It is formed when water at times of flood flows over the adjacent land surface carrying debris and Earth and depositing it in an alluvial fan of fan-shaped deposits. This phenomenon commonly occurs in concentric rings around a major stream or river that is concentrated near its center because of narrowing at both banks. In other instances, flood plains are formed by erosion on a flood's path, causing the relatively soft and deep soil to erode and form an alluvial plain which may be quite wide, very flat and level, or somewhat uneven with river terraces. Ideally, flood plains have an elevation that is within one meter of the current water level. A flood plain can be as small as land around and including a single home, apartment building or school. In other instances, the flood plain can be very large area indeed (such as Bangkok).

Fluid

Fluid is one of the three fundamental states of matter, along with solid and gas. They are substances that flow and take the shape of a container. The only state of matter having a definite volume but taking the shape of its container is the solid-state. On Earth, where there is gravity, fluids move downward toward the lowest possible level. Therefore, it is most likely that water flows downward in a container and lava flows upward in a container. This occurred when Earth was young and magma came close to the surface of the planet. Fluid is one of the four fundamental states of matter (the other three being solid, gas, and plasma). A fluid has a definite volume, but not a fixed shape; it flows without viscosity, however, unlike gases, it does not disperse to fill every space. Fluids were long considered to be a single continuum (the term hydrodynamics was coined by Horace Lamb in 1871 to describe the hydrostatic paradox). Since then, it has been discovered that fluids can be divided into several phases (such as gases, liquids, and solids) using such properties as specific volume or chemical composition (Goldstein et al., 2017). Fluid state can be recognized by its ability to deformation under shear stress in a measurable way. The texture of a fluid can be characterized as either being continuous or discrete. Fluids are a crucial part of many physical phenomena, including blood flow (Figure 59).

Figure 59. Fluid flowing.

Source: Picture by NASA.

Fluorine

Carbon, nitrogen, oxygen, and fluorine are the only elements that are typically found in biological molecules. Fluorine has a high electronegativity value; it is 17.5 on the Pauling scale. Florin in basic compounds is a strong electron acceptor and has a high ability to attract electrons from other atoms. Of the elements present in biological molecules, fluorine is the most reactive one. This chemical element has the symbol F, and an atomic number of 9. It is the lightest halogen and exists as a highly toxic pale yellow diatomic gas, at standard temperature and pressure. Conditions of normal usage ensure that it remains in gaseous form, and presents little threat of biopersistence or bioaccumulation due to its very high rate of decomposition, making it one of the most readily available chemicals for use as a biocide (Escorihuela, 2018). At standard pressure and temperature, fluorine forms diatomic fluoride in all proportions of matter. Fluorine is a ubiquitous component of the environment, even in dilute form, much like other halogens. Despite its toxicity, many organisms depend on fluorine compounds for normal growth. Fluorine is a naturally occurring chemical element that is most widely used in the synthesis of industrial products, including pharmaceuticals and refrigerants. It is obtained from fluorspar, fluorite (calcium fluoride), or cryolite (Figure 60).

Figure 60. Fluorine.

Source: Picture by Live Science.

Fossil

The Earth formed about 4.5 billion years ago, and though the planet itself will long outlive us all, there is evidence that humanity may be nearing its end. It humbles us to reflect on the passing of time, on the blue marble we call home, and on our place in the cosmos. This is why we do what we do – to craft great watches that help you build a lasting relationship with time. Geology is the science that studies the solid and liquid parts of the

Earth and its history. Geologists study rocks and fossils—objects made by living things, which died long ago. They use the findings to learn about past climates and environments that have changed over time. Fossils can tell where certain plants and animals lived many years ago. There is still a lot researchers do not know about our planet's history. The word 'fossil' has two meanings (Duffy, 2014). Fossil fuels are derived from the fossilized remains of dead organisms. And in its original geological use, the word 'fossil' referred to any remains, trace, or impression from past life. In what is now known as paleontology, the science dealing with prehistoric life, fossils are the remains of once-living things of an earlier geologic age found within sedimentary deposits. Fossils are usually made of stone, but soft and fleshy ones have been found made of amber, coral, and even blood. A bone found in a coal mine may be called a fossil, but this is not technically correct. In general, fossils are the remains of organisms that lived long ago and were buried quickly under natural circumstances, for example, by deposition of sediments (Figure 61).

Figure 61. Fossils.

Source: Picture by Nature.

Fracture

Fracture is a natural process, the physical result of applied stress. In geology and Earth science generally, fracture is any separation in a geologic formation, or in an object made of rock. Fracture strength or breaking strength is defined as the stress that causes a solid material to break in two. The surface of the Earth, parts of which are broken into pieces by geological processes.

These processes include faulting and the movement of tectonic plates. A fracture is a zone where the Earth's crust has been fractured in an earthquake or some other event. Movement occurs along planes of weakness (fracture planes) within rock or soil, such as existing joints or zones of weakness in sedimentary material (Clowes, 2010). Fractures are a fundamental part of the Earth's crust and can be separated into three main types: intraplate fractures, interplate fractures, and intracontinental fractures. Intraplate fractures are caused by strain without a change in stress. This type is seen in Wyoming and Montana. Interplate fractures are formed during the process of plates colliding, sliding past, or subducting each other. Intracontinental fractures occur when two continents collide (Figure 62).

Figure 62. Fracture.

Source: Picture by Sciencing.

Friction

Earth's surface is a dynamic mosaic of tectonic plates in constant movement. Plates collide, diverge, and transform into continents and ocean basins. Even as the continents themselves move above the Earth's mantle, friction is generated between Earth's outermost layer—the lithosphere—and underlying hot mantle rocks, to move entire continents at rates up to several inches per year. Friction is the force resisting two different materials that slide or otherwise touch each other. Friction occurs when two surfaces are moving past one another and one surface exerts a force upon the other (Clary and Wandersee, 2011). Nonslip synthetic rubber material helps reduce slippage on smooth surfaces, such as sport courts, pool decks, park floors, patios, and poolside decks. Also reduces possible injury from falls on slick surfaces in public areas. Friction is the resistance that acts to oppose motion between two surfaces in contact. It can be thought of as a force that pulls objects apart from one another. It is a very important aspect of Earth science and is responsible for the formation of many geological features such as mountains, valleys, and volcanoes. Climatic conditions, such as frost or ice, can also cause friction.

Fuel

Fuel is a substance that is burned to provide energy to drive machines or heat buildings. It comes from the Latin word for wood: " fuere," meaning something that readily burns. Earth's main source of fuel is its supply of coal, which is formed from decayed plants and trees.

Figure 63. Fuel extraction.

Source: Picture by Britannica.

As these plants grow, they absorb carbon dioxide (CO_2) from the air through their leaves. This decayed matter eventually goes into layers as limestone. The pressure of the overlying layers form coal. Coal has been used by humans for heating and cooking for hundreds of years (Cossu et al., 2010). It is mined from the Earth in different parts of the world. Because fossil fuels take millions of years to form. It would be wrong to make the Earth go through continuous extraction in the name of fuel. Earth is essentially going to give us finite, but clean sources of energy. Fossil fuels are the most reliable source of energy on Earth. This includes coal, oil, and natural gas. Fossil fuels are made from the remains of plants and animals that died millions of years ago (Figure 63).

Fumarole

Fumaroles are openings in the crust of the Earth that emit gases. In contrast to other gas-emitting springs such as geysers, fumaroles are characterized by gas that is emitted from the Earth at very low to moderate temperatures and with few or no minerals dissolved in it. They range greatly in size, from tiny cracks to immense craters.

Figure 64. Fumarole.

Source: PixCove.

The gases rising through the fumarole give off heat and often make a roaring sound if the fumarole is quite large. At a fumarole, steam, gases, and volcanic material such as volcanic glass, interact at the surface of the Earth and emit light before they return to the atmosphere. Fumaroles are volcanic vents that emit steam, gases, and hot springs. These surfaces can reach up to 500°F. Fumaroles may be identified by a surface mound that emits a vapor or gas (Cerase, 2019). The mound is caused by boiling water beneath the surface. They result from volcanic activity; hot magma rises through a crack in the Earth's surface and comes into contact with rocks, creating heat and water. When the gases cool, they produce large quantities of sulfur compounds. These sulfur compounds then condense and harden as they hit cooler air, creating a column of ash which resembles a pelican's beak (Figure 64).

G

Galaxy

Galaxy is the term used to refer to a system of dark matter, dust, interstellar gas, stellar remnants, and stars bounded gravitationally. The Earth's galaxy is called the milky way. There are different galaxies having different sizes ranging from dwarfs with a few 100 million stars to giant galaxies with 100 trillion stars. The stars orbit its galaxy's center mass. A galaxy's visual morphology is used to categorize it. According to the visual morphology a galaxy can be categorized as irregular, spiral or elliptical. There are galaxies said to have supermassive black holes at the centers. For Earth's galaxy (milky way), the central black hole is called the Sagittarius A. Its mass is 4 million times greater than the Sun. Data obtained from march 2016 shows that the oldest and most distant galaxy observed is the GN-z11. The GN-Z11 has a comoving distance of 32 billion light-years from Earth. It is estimated that it has been in existence 400 million years after the big bang (Britner, 2008). Between galaxies, there is space. The space is composed of a tenuous gas or an intergalactic medium having an average density of less than one atom per cubic meter. Galaxies can be organized into superclusters, clusters, and groups based on its gravity. The Andromeda galaxy and the milky way are part of the local group. They fall under the Virgo supercluster. The associations can be formed on a small scale or large scale. In the large scale, the associations are grouped into filaments and sheets surrounded by immense voids. The Virgo supercluster and the local group are within a larger cosmic structure called Laniakea (Figure 65).

Figure 65. There are a number of galaxies in the universe.

Source: https://www.nasa.gov/mission_pages/chandra/images/chandra-sees-evidence-for-possible-planet-in-another-galaxy.html.

Gamma

Gamma is a term used in referring to cosmic rays. Gamma rays form part of the electromagnetic spectrum. They are characterized by a large amount of energy and relatively small wavelengths. Gamma rays can be produced by very hot and energetic objects in the universe. These objects include supernova explosions, regions in blacks holes, pulsars, and neutron start. Gamma rays can also be produced on Earth through lighting and nuclear explosions. There are various tools used in detecting the presence of gamma rays. Compared to other cosmic rays, gamma rays cannot be reflected or captured by mirrors (Butler, 2016). As they have relatively short wavelengths, gamma rays can pass through space within the atoms of a detector. For this reason, densely packed crystal blocks are used in gamma-ray detectors. As the rays pass through the detector, they hit electrons in the crystal. The process of gamma rays colliding with electrons is called Compton scattering. As the gamma-ray hits an electron, it loses energy. The collisions result in the generation of charged particles that can be detected by a sensor (Figure 66).

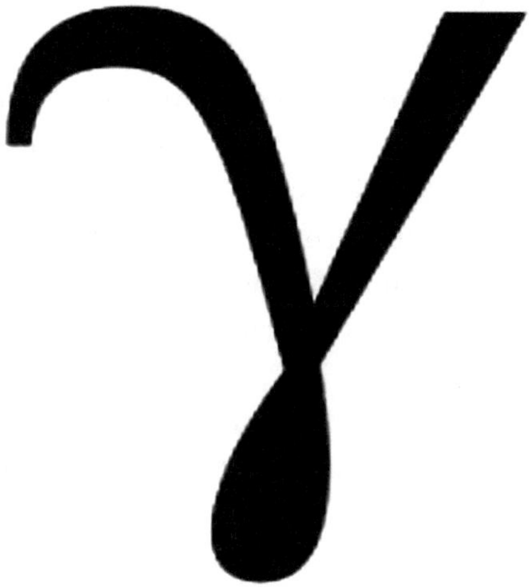

Figure 66. The gamma symbol.

Source: https://www.shutterstock.com/search/gamma+symbol.

Gaseous

Gaseous is a term used to refer to the state of the Earth's atmosphere. The Earth's surface is described as a layer of gases held closed to the Earth by gravity. These gases play a vital role in supporting animal and plant life. It plays a crucial role in maintaining a global average surface temperature of about 14°C. The gaseous Earth's surface determines climate and weather. It screens south harmful radiation produced by the Sun and they act as a buffer against meteorites. As a gaseous system, the Earth's atmosphere is made up a variety of gases. The main components of the atmosphere include oxygen, carbon dioxide (CO_2), argon, and nitrogen. There is also a small percentage composed of trace amounts of other gases (Bohle, 2015). A small percentage of the Earth's gaseous atmosphere is made up of water vapor that takes about 0–4% of the Earth's atmosphere. The variation of water vapor depends of the location, pressure, and temperature. Of the gaseous composition of the Earth's atmosphere gases such as CO_2, oxygen, and nitrogen play a vital role in human life. Oxygen is taken in by living life forms and exchanged between it and the atmosphere. Oxygen is formed through respiration and photosynthesis. Nitrogen is removed from the atmosphere by some bacteria. These bacteria fix nitrogen into the soil, allowing plants to take up nitrogen and use it in the production of proteins. CO_2 is used in photosynthesis and is produced during respiration.

Geocentric

Geocentric is a term used to refer to a system in which the Earth is regarded to be the center of the solar system. Basic information of the solar system is that the star, Sun, planets, and the moon rotate around the Earth. The Earth is at a fixed position and rotates around its orbit. These bodies move around the Sun with a uniform circular motion. A geocentric system is said to be unchanging and ethereal. By considering the Earth as the center of a solar system, the geocentric theory was developed. The theory postulated that the Earth was the center of the universe (Bohle, 2015). For early astronomers, this theory was very evident as the stars, planets, moon, and Sun moved about a stable motionless earth. These bodies move round the Sun along a circular path every day. Although the theory seemed to be a reasonable assumption, it was rejected by modern science. The geocentric theory was supported by the fact that objects fell towards the Earth (Figure 67).

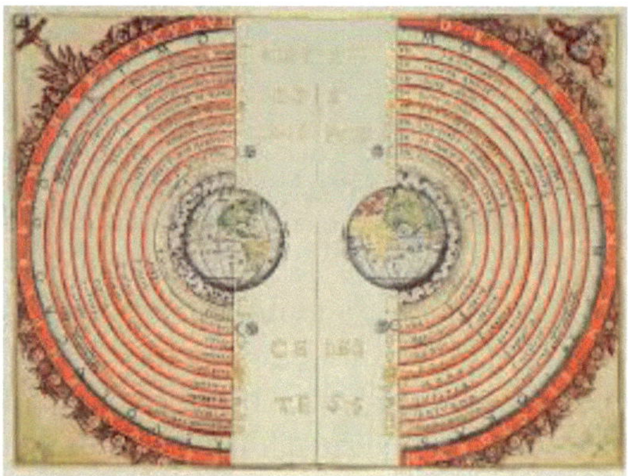

Figure 67. The geocentric model.

Source: https://en.m.wikipedia.org/wiki/Geocentric_model.

Geologic

Geologic is a term used to refer to the scientific study of the origin, structure, and history of the Earth. It is based on geology. Geologic is concerned with the solid and liquid Earth, its rock composition and the processes followed as they gradually change with time. Geologic can also be associated with the study of natural satellites and solid features of a terrestrial planet. Geologic can also be used to refer to events, hazards, and time (Zhuang et al., 2019).

Geologic events are events that resulted in geologic processes which act to modify geologic entities. Geologic hazards are adverse geologic conditions capable of causing loss of life or property or damage. Geologic time scale is a chronological dating system used in the classification of the geologic strata in time.

Geologic Time

Geologic time refers to an extensive time interval occupied by the geologic history of Earth. Formally the geologic time begins at the start of the Archean Eon and is used in the current day and age. When discussion geologic time, the term geologic time scale comes into the context. The geologic time scale is a chronological dating system used in the classification of a geologic strata in time. The modern geologic time scale includes the Hadaen Eon (Yu and Gong, 2012). This time scale is an informal interval which extends from about 4.6 billion years ago to 4.0 billion years ago.

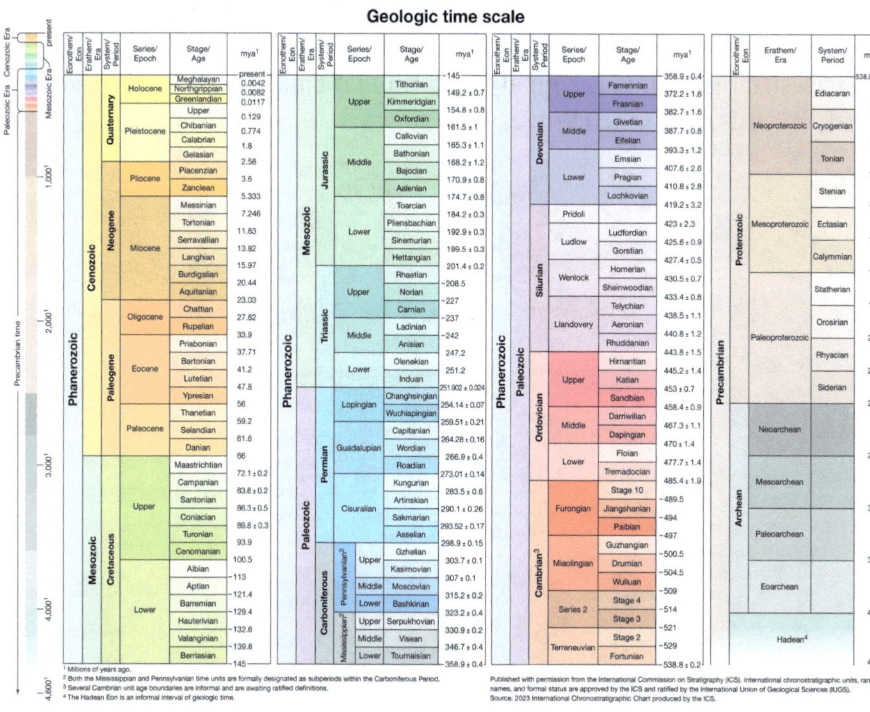

Figure 68. There are different geological time calendars.

Source: https://www.britannica.com/science/geologic-time.

A segment of Earth's history represented by and recorded in the planet's rock strata is the geologic time. Therefore, the geologic time scale can be defined as a calendar for events in Earth history. The geologic time scale subdivides time into named units of abstract time. This abstract time is called ages, epochs, periods, eras, and eons. Stratigraphy is used in the enumeration of the geologic time units. Stratigraphy is the correlation and classification of rock strata. Within rocks, there are fossil forms which provide the chief means of establishing geologic time scale. It is achieved by timing the disappearance and emergence of widespread species from fossil records used in the delineation of the endings and beginnings of periods, epochs, ages, and other intervals (Figure 68).

Geological

Geological is a term used to refer to the scientific study of the origin, structure, and history of the Earth. It is based on geology. Geological systems are concerned with the solid and liquid Earth, its rock composition and the processes followed as they gradually change with time. Geological can also be associated with the study of natural satellites and solid features of a terrestrial planet. Geological is a term that can also be used to refer to events, hazards, and time (Witze, 2013).

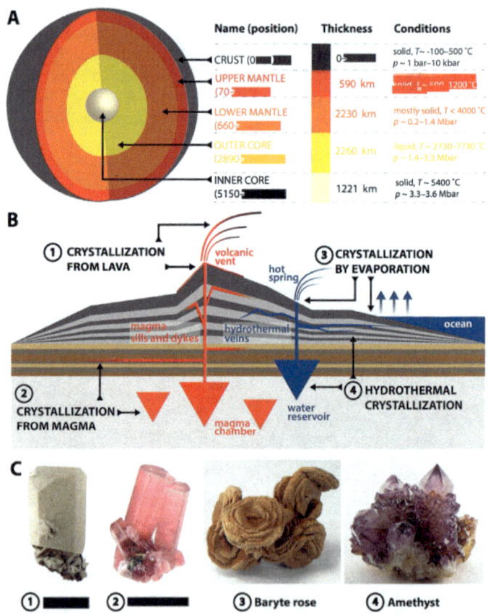

Figure 69. There are various kinds of geological materials.

Source: https://www.researchgate.net/figure/Geological-materials-formation-conditions-and-processes-A-Structure-of-the-Earth-with_fig3_317273677.

Geological events are events that resulted in geologic processes which act to modify geologic entities. Geological hazards are adverse geologic conditions capable of causing loss of life or property or damage. Geological time scale is a chronological dating system used in the classification of the geologic strata in time. There are geological materials such as rock, magma, and minerals (Figure 69).

Geologist

A geologist is an individual who specializes in the study of gaseous matter, liquid, and solids that constitute the Earth and other terrestrial planets. They also study the processes that result in the formation of matter. For an individual to qualify as a geologist, they study geology and have backgrounds in sciences, biology, chemistry, and physics. A very important component in geology is field research. There are various sub-disciplines that incorporate digitalized work and laboratory. There are various departments under geology. There are geologists who work in the mining and energy sector (Wilkinson et al., 2020). They mostly search for natural resources such as base metals, natural gas, and petroleum.

Figure 70. There are different types of geologists.

Source: https://www.bsu.edu/academics/collegesanddepartments/environment-geology-natural-resources/academic-programs/bachelors/geology.

They also work towards the prevention and mitigation of damage from disasters and natural hazards such as landslides, tsunami, volcanoes, and earthquakes. Results from their studies have been used in warning the public of the occurrence of the events. They are also actively involved when making decisions regarding climate change. There are various kinds of geologists. They include environmental geologists. They work towards finding solutions to various environmental issues. This is achieved by investigating and observing natural hazards, pollution, earthquakes, erosion, and flooding. There are marine geologists. Marine geologists study the processes of landforms beneath the ocean (Figure 70).

Geology

Geology, also known as Earth science or geoscience is a term used to refer to the study of the Earth, its material composition, the structure, not its materials and processes acting upon them. Geology entails the study of organisms that have inhabited Earth over the years. A very important part of geology is the study of Earth's materials, processes, organisms, and structures and changes they have undergone over the years. Geology tends to overlap other Earth sciences such as atmospheric sciences and hydrology. For this reason, it is related as a major aspect of integrated Earth system science and planetary science. As a study, geology seeks to understand the structure of the Earth (above and beneath its surface) and the processes that have shaped that structure. It has been useful in providing tools used in determining the absolute and relative ages of rocks found in given locations (Wallace-Wells, 2019). They are useful in describing the history of the rocks.

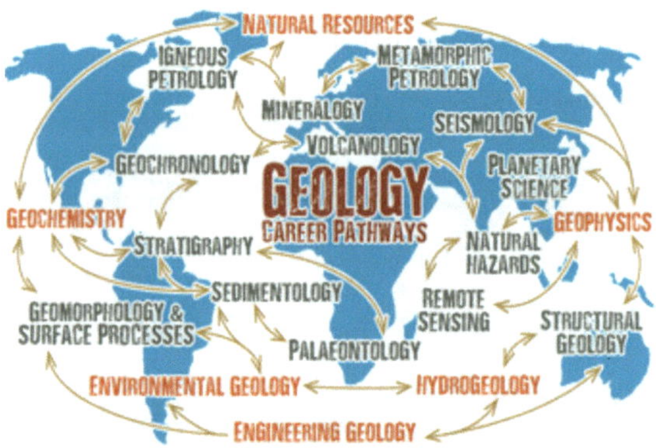

Figure 71. Geology is the study of the Earth and its structure.

Source: https://www.geolsoc.org.uk/Geology-Career-Pathways/What-is-Geology.

Geologists are able to combine these tools to enable them to chronicle the geological history of the Earth as a whole and also demonstrate the age of the Earth. Geology is useful in providing primary evidence of Earth's past climates, evolutionary history of life and plate tectonics. There are different methods that geologists can use to understand the Earth's structure and evolution. They include numerical modeling, physical experiments, chemic analysis, geophysical techniques, rock description and fieldwork. Hydrocarbon and mineral exploitation and exploration involves geology. Geology is also important in understanding natural hazards, providing insights into past climate change, remediation of environmental problems and evaluating water resources. As an academic discipline, geology is central to geological engineering and is actively used in geotechnical engineering (Figure 71).

Geothermal

Geothermal is a form of thermal energy in the Earth's crust. Geothermal energy originates from radioactive decay of materials and formation of planets. Due to the high pressure and temperature in the Earth's interior causes some rocks to melt while solid mantle behaves plastically. This causes parts of the mantle to convect upwards since it is lighter than surrounding rocks while temperatures at the core-mantle boundaries can reach about 7,200°F (Figure 72) (Toth et al., 2014).

Figure 72. Geothermal energy is used in the generation of electricity.

Source: https://www.thinkgeoenergy.com/geothermal/what-is-geothermal-energy/?amp.

There are different sources of heat deep in the Earth. Some of the heat is a result of heat decay from radioactive elements, frictional heating, and heat from the formation of the planet and accreted that has been lost as heat from the Earth takes a longer time to be released from the Earth. Transfer of heat from within may be due to convective transport of heat within solid mantle or liquid outer core of the Earth. Heat is also lost through slower conductive transport of heat. Temperature values from the Earth's surface tends to increase as temperature is measured deeper. This is due to geothermal gradient. Increase in temperature per 100 m of depth is highlighted by the geothermal gradient (Tollefson, 2015). One of the main forms of generated electrical power is geothermal power generated from geothermal energy. Certain technologies are used in geothermal power stations. They include binary cycle power stations, flash steam power stations and dry steam power stations. About 26 countries use geothermal electricity while 70 countries use geothermal heating. Among the advantages of geothermal power is a sustainable, renewable energy source. This is attributed to the fact that heat extraction is smaller compared to the Earth's heat content.

Glacial

Glacial is the term used to denote the presence of ice more so glaciers. The term glacial is also used when referring to landforms hence the name glacial landforms. Glacial landforms are any products of meltwater and flowing ice. Such kinds of landforms are produced in glaciated areas. During the course of history, large expansion of present-day glaciers has recurred. The last ice age that ended about 20,000 to 15,000 years ago, ice-covered about 30% of the land surface. Glacial landforms may continue to exist if they are not obliterated by landscape-modifying processes. There are areas devoid of glaciers but were once glaciated (Thessen, 2016). The term glacial could also be used to refer to a period hence the name glacial period which is a 1,000-year interval within an ice age naked by glacier advances and cold temperatures. The term inter-glacial refers to a period of water climate between glacial periods. Records show that the last glacial period ended 15,000 years ago. The current inter-glacial is the Holocene. A time period in which there are no glaciers in the greenhouse climate state. Therefore, there are different kinds of glacial periods including the quaternary period and penultimate glacial period.

Glacier

Glacier is a term used to refer to a large perennial accumulation of crystalline sediment, rock, snow, and ice that originates on land and finds its way down the slope. Movement down the slope is influenced by gravity and the weight of water. Glaciers may form or exist in areas where winter precipitation produces significant accumulation of snow, mean annual temperature is close to the freezing point and temperature experience in the region throughout the year does not cause complete loss of previous snow accumulation. As the accumulation of snow continues it results in the presence of a large mass of snow allowing the start of the metamorphism of snow to glacier. Some of the factors used in the classification of snow is the thermal regime, location, and their size. Glaciers are studies and give much information of climate change. There are different types of glaciers characterized by their behavior, thermal characteristics, and morphology. The slopes and crests of mountains are formed by alpine glaciers. Valley glacier are those that fill valleys. Alpine glaciers are also called mountain glaciers. Ice field or ice cap is a term for a large body of glacial ice stride. By definition ice caps have an area less than 50,000 km^2. Ice sheets are glacial bodies larger than 50,000 km^2 (Telling et al., 2017). They are also called continental glaciers. They are normally several kilometers deep causing the obscurement of underlying topography. There are two extant ice sheets found in Greenland and Antarctica. These ice sheets contain large quantities of freshwater. Ice shelves are portions of ice caps or sheets that extend into water. Ice shelves are normally thin with reduced velocities and limited slopes. Ice streams are narrow, fast-moving sections of ice sheets. In some cases, the ice streams drain into ice shelves while others drain directly into the sea. Examples of glaciers that terminate in the sea include tidewater glaciers (Figure 73).

Figure 73. Global warming has caused the rapid decline in glaciers.

Source: https://www.google.com/amp/s/www.nbcnews.com/news/amp/rcna791.

Global

Global is a term used when relating to the whole world. It is used alongside other words when referring to things affecting the whole Earth in general. For instance, there term global warming is used to refer to climatic problems affecting the whole world in general. Global warming is as a result of air pollution globally therefore measures taken to stop global warming are enforced globally (Showstack, 2008Global changes could include changes in temperature, rainfall, and vegetation. Global environment is a term used to refer to the state of the environment worldwide. It focuses on the state of vegetation, wildlife, and climate at a global scale (Figure 74).

Figure 74. Global refers to a worldwide scale.

Source: https://www.entrepreneur.com/article/226554.

Globe

A globe is an accurate representation of the Earth's curved surface. In most cases, globes have a geographic coordinate system and a scale. The length of the arc is the shortest distance between two points that connects the two points in the globe.

Figure 75. A globe is a representation of the Earth.

Source: https://www.ubuy.ke/en/catalog/product/view/id/34661364/s/wizdar-8-world-globe-for-kids-learning-diy-assemble-educational-rotating-world-map-globes-large-size-decorative-earth-children-globe-for-classroom-geography-teaching-desk-office-decoration.

A terrestrial globe is a model globe of the Earth while a celestial globe is a model globe of the celestial sphere. Globes are useful in giving details of its subject. Terrestrial gloves show water bodies and landmasses. It may so show a network of longitude and latitude lines, major cities, and nations (Schmitz, 2011). In some globes, relief has been raised to indicate the presence of large landforms and mountains. Celestial globes are more advanced as they shoe positions of other prominent astronomical objects and notable stars. Celestial globes divide the celestial sphere into constellations and can therefore show apparent positions of stars in the sky. Celestial globes omit planets, moon, and Sun but indicate the ecliptic along which the Sun moved (Figure 75).

Gravitational

Gravitational is a term used when relating to the movement of an object toward the center of gravity. The Earth is a center of gravity. This is why when objects fall, they fall downwards. Objects move towards the Earth as they are attracted by a force called the gravitational force. The gravitational force pulls things towards the center of the Earth. Different planets have different gravitational forces with Jupiter having the highest gravitational force (Schmitz, 2012). Gravitational force does it determine the internal properties of matter however it has a long reach universal action allowing it to control the trajectory of bodies in the solar system. Gravitational force is mostly influenced by the weight of the object. Objects have a downward force of gravity or weight which is proportional to their mass. The Earth's mass exerts a downward force on the object. Gravitational force is measured by the acceleration of the freely falling objects. Heavier objects tend to move faster towards the center of the Earth faster than lighter objects (Figure 76).

Figure 76. Gravitational force causes objects to move towards the Earth.

Source: https://tuitionphysics.com/nov-2017/what-is-gravitational-field-of-a-point-mass/.

Gravitational Field

Gravitational field is a model used in explaining the influences that a massive body extends into the space around itself. The field produces a force on another massive body. Gravitational phenomena can be explained by a gravitational field. The gravitational field can be measured in newtons per kilogram. To better understand gravity, Pierre-Simon Laplace and Isaac Newton attempted to model gravity by likening it to a fluid or radiation field this resulted in gravity being explained in terms of a field model (Rao, 2012). When using the model, particles are seen to distort space-time via their mass rather than attracting each other. The force is measured and perceived to be the distortion. From the model, it is deduced that matter moves in a certain way as a response to the curvature of space-time. It also states that there is no gravitational force. Gravity is different from other forces in that it obeys the equivalence principle. Gravitational field around an object can be shown in two ways. They include field lines and arrows. The magnitude of the field lines is shown by the arrows. The arrows also indicate the direction of forces. The direction in which force would act on an object placed at a point is indicated by the field lines (Figure 77).

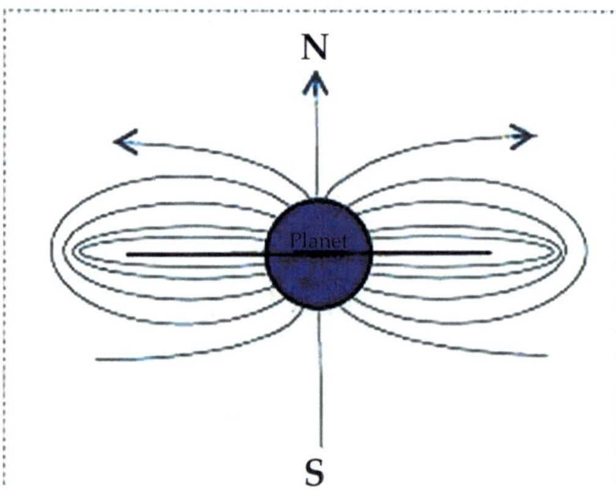

Figure 77. Gravitational field are characterized by gravitational field lines.

Source: https://www.researchgate.net/figure/The-direction-of-the-gravitational-field-The-poles-are-distinguished-with-the-black-line_fig7_279698263.

Gravity

Gravity is a force of attraction that acts between matter. In mechanics, gravity is also called gravitation. Compared to other forces in nature, gravity is the weakest and is not used in determining the internal properties of matter.

Gravity has a long reach universal action allowing it to control the trajectory of bodies in the solar system. Gravity is influenced by the mass of an object. All objects have a weight that is proportional to their mass. The mass of the Earth exerts a force of gravity on objects (Rivet and Kastens, 2012). Acceleration caused by gravity on a free-falling object is used to measure gravity. Acceleration of gravity is different in different surfaces. Acceleration of gravity in the moon is very small compared to Earth as the moon has a small gravitational force. For this reason, objects tend to float in the moon rather than moving towards its center. However, in Jupiter, the acceleration of gravity is very high such that object burn out before it reaches its surface.

Ground Level

Ground-level can also be used to refer to ozone hence the name ground-level ozone. The ground level ozone is an ozone layer created by chemical reactions between volatile organic compounds and nitrogen. Most of these gases are not directly emitted into the air. These gases find their way to the ozone through chemical plants, refineries, industrial boilers, power plants or cars. Ground-level ozone is highly likely to reach unhealthy level during cold and hot months. This layer of ozone can also be transported over long distances by wind.

Groundwater

Groundwater is termed as the Earth's largest reservoir. Groundwater is found in sediment, porous rock, and aquifers. Movement of water into porous media is described by capillary action. Water is attracted to soil particles causing its accumulation. Through capillary action, water find its way from wet soil to dry soil. Aquifers can be found at different depths with most aquifers being deep below land surface and very few being just below the Earth's surface. Several aquifers can be found in a particular region. Some aquifers are found in deserts. A mountain area can be the source region for an aquifer for a desert though it may be far away (Pyle et al., 2018). There are a number of factors influencing the amount of water available to enter groundwater. They include water retention, land use in the area, the vegetation cover, the type of rock found at that surface, slope of the land and local climate. More groundwater is found in regions where soil is exposed, the land of flat and there is a lot of rain. Water in groundwater aquifer could have stayed there for minutes or even years. For groundwater that has stayed for a very long duration of time, it is referred to as fossil water more so those that have stayed since the end of the ice age. The amount of recharge should be equal to the amount of discharge for groundwater aquifer to contain the same amount of water. Groundwater is very valuable and is a major source of water. As a source of water, groundwater is renewable and is sustainable when water pumped out is replenished. The type of pump used depends on the depth of groundwater (Figure 78).

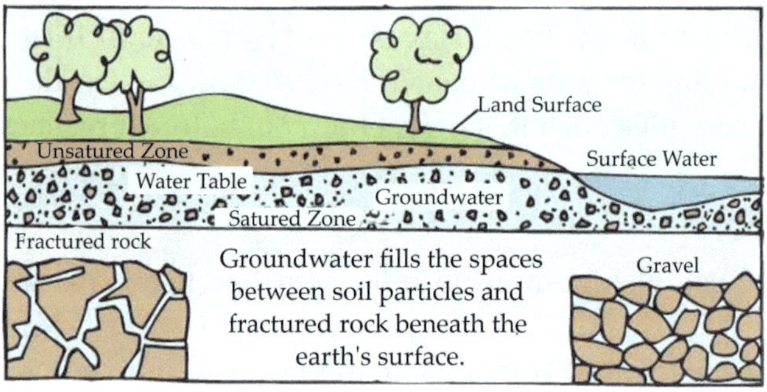

Figure 78. Water finds its way underwater through various ways.

Source: https://www.groundwater.org/get-informed/basics/groundwater.html.

H

Heat Energy

Planet Earth is made in such a way that it contains heat within it. The heat energy within the Earth's structure is referred to as geothermal energy. In most parts of the world, geothermal energy is being utilized for different activities, for it is a renewable source of energy since the Earth is continuously producing heat from within (Figure 79).

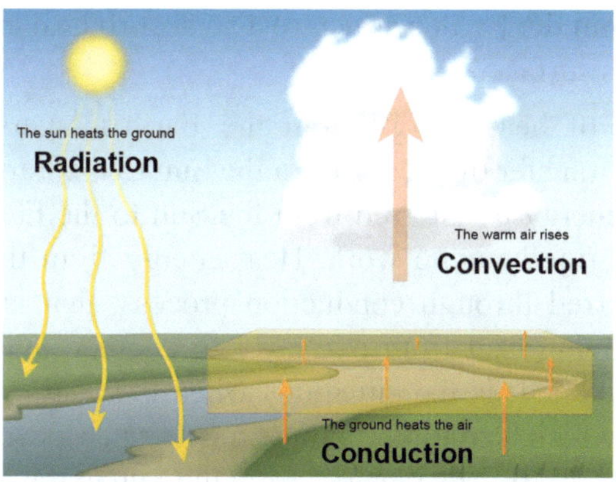

Figure 79. The transfer of heat energy from the Sun to the Earth's surface and vice versa.

Source: https://www.google.com/imgres?imgurl=https%3A%2F%2Fscied.ucar.edu%2Fsites%2Fdefault%2Ffiles%2Fimages%2Flarge_image_for_image_content%2Fheat_transfer_large_0.png&imgrefurl=https%3A%2F%2Fscied.ucar.edu%2Flearning-zone%2FEarth-system%2Fconduction&tbnid=HDD8hmt9uKEyaM&vet=12ahUKEwiH-8mK3N70AhUWphQKHT92A8gQMygAegUIARCrAQ.i&docid=Hq6TN1x2uJNofM&w=768&h=576&q=heat%20

energy%20in%20Earth%20science&ved=2ahUKEwiH-8mK3N70AhUW-phQKHT92A8gQMygAegUIARCrAQ.

Heat energy within the Earth's structure is produced by the radioactive particles as they undergo the decaying process. The radioactive particles are in all types of rocks to be found on the Earth's surface. Generally, the Earth is made up of different layers, which are the main sources of heat energy. The inner most layer is the inner core, which is made up of solid iron. The inner core has a temperature of over 11,000°F, a higher temperature than that of the Sun (Press, 2008). The second innermost layer, after the inner core is the hot molten magma, with temperatures of approximately 7,000°F. The third from the bottom of the Earth's surface is also magma, however, it is no longer in molten form, rather in solid form. It has a temperature of approximately 400°F. The outer most layer of the Earth's surface is the crust. Crust is the layer that makes up the Earth we see with our eyes, the soil, the continents, ocean floors, river floors among others. The crust is mainly a solid rock, which provides the foundation of the ground we establish different structures (Plotnick et al., 2014). The visible rocks on the Earth surface absorb the heat from the third layer of the Earth's surface, controlling the heat energy experienced on the Earth's surface. Different water bodies within the Earth's structure also absorb heat energy from deep within the Earth's surface.

The transfer of heat energy from the Earth's layers could also be equivalent to the transfer of energy from the Sun to the Earth's surface. The total amount of energy transferred from the Sun to the Earth's surface can be equated to thermodynamic work. Heat energy from the Earth's layers is mostly transferred through conduction process; that is, direct conduct of the layers, moving heat from one layer to another and finally onto the outer layer, the crust. It is the situations when the heat within the Earth's layers is so high which lead to Earth movements, such as volcanic eruptions (Pennington et al., 2020). Heat energy from the Sun to the Earth's surface is transferred through radiation, since both bodies are greatly separated from each other. The ozone layer is known to help reduce the amount of heat energy the Earth's surface receives from the Sun, by filtering out the Sun rays which will be harmful if they were to reach the Earth's surface.

Heavenly Body

Heavenly body, is a term used to refer to any natural object in space, be it a planet, star, moon, and the Sun, among others. The heavenly bodies revolving around fellow heavenly bodies are referred to as their satellite; however, it is referred to as a satellite, if, and only if, the other heavenly body it is revolving around is a planet. For instance, the moon can be referred to as a satellite to planet Earth (Figure 80) (Perdigão et al., 2020).

Figure 80. An image of heavenly bodies.

Source: https://www.google.com/imgres?imgurl=https%3A%2F%2Fpediaa.com%2Fwp-content%2Fuploads%2F2021%2F03%2FDifference-Between-Celestial-and-Heavenly-Bodies_1.jpg&imgrefurl=https%3A%2F%2Fpediaa.com%2Fwhat-is-the-difference-between-celestial-and-heavenly-bodies%2F&tbnid=7d6pgXEh3LXoyM&vet=12ahUKEwieocfM2970AhXGDWMBHQLHC-AQMygBegUIARCvAQ.i&docid=GrB9IIqcIFfMWM&w=1024&h=576&itg=1&q=heavenly%20bodies%20in%20Earth%20science&ved=2ahUKEwieocfM2970AhXGDWMBHQLHC-AQMygBegUIARCvAQ.

A star is believed to be a circular accumulation of gases with the ability to produce their own light. Stars do also have the ability to produce heat energy by converting one of their main gases such as Hydrogen gas into Helium gas which is categorized under hot gases. From the Earth's surface, stars are seen as tiny lights, however, they are big in size and experience a high gravitational attraction within themselves and other heavenly bodies around them. The Sun is also categorized under star, with a gigantic size, and gives planet Earth the needed energy making life easier for human beings (Parkes and Morris, 2001).

Planets are said to be spherical objects which satellites to the Sun for they revolve around it. Each planet has its own pathway in space, referred to as an orbit, allowing them to revolve around the Earth without colliding with each other. Planets could either be made of rocks, metals, or gases such as hydrogen, nitrogen, and methane. Earth is the only planet known to support life; however, there are ongoing experiments to prove that planet Mars could also support life in the future. There are eight known planets which include; Mercury, Venus, Earth, Mars, Jupiter, Saturn, Uranus, and Neptune.

Satellites, as stated earlier, are heavenly bodies which revolve around planets. Satellites could either be natural or artificial. Natural satellites include the likes of the moon to the Earth, which is strongly bound to the Earth, by the Earth's gravitational pull. Artificial satellites are made by astronauts with the interest of studying the other planets in space (Özyurt et al., 2021).

Comets are heavenly bodies which are made up of frozen water and rocks. The fall out of outer edges of the planets. In situations where they revolve close to the Sun, the frozen water in them melts, creating a tail-like trail in them.

Asteroids are rocks either made up of metal or minerals which revolve around the Sun. Asteroids are mostly found in between planet Mars and planet Jupiter; hence this region is referred to as an asteroid belt.

Meteors are heavenly bodies which are pulled out of space into the Earth's atmosphere by the strong Earth gravitational force. In most cases, meteors burn up as they enter the Earth's atmosphere, creating a unique pattern as if a star has fallen, hence the name shooting stars. In some rare situations, the meteors are big enough and they manage to reach the Earth's ground creating a crater. Such meteors are referred to as meteorites.

Galaxy is a term used to refer to a group collection of heavenly bodies, being held closer to each other by gravity. Most galaxies are made of stars; however, the Sun and the planets are part of the Milky Way galaxy (Orion, 2019). Galaxies close to the Earth are clearly visible from the Earth's surface on a clear night; however, galaxies, far away from the Earth are confused to be stars.

Hemisphere

Hemisphere, as the name suggests, is a term used to refer to the part which is half of a sphere. Planet Earth being spherical in nature, is divided into two equal parts known as the hemispheres. However, the Earth is considered to have four hemispheres, which are: The Northern hemisphere, The Southern hemisphere, The Eastern hemisphere, and The Western hemisphere.

The main reason as to why the Earth is said to have four hemispheres is because it divided both vertically and horizontally by longitudes and latitudes, respectively. The main latitude, the equator, divides the Earth into the Northern and the Southern hemisphere; whereas, the main longitude, the Prime Meridian, divides the Earth into either the Eastern hemisphere of the Western hemisphere (Oreskes, 2015).

The southern hemisphere is made of five different continents, which include Antarctica, Africa, Australia, part South America and a small part Asia. It is also made up of Southern Ocean, Indian Ocean, South Atlantic Ocean, and South Pacific Ocean. The Northern hemisphere is made up of Europe, North America, a large part of Asia and a quarter of South America. Both the southern hemisphere and the northern hemisphere experience different climatic conditions. The northern and southern hemisphere experience different temperatures at the time of the year; for instance, when it is winter season in the southern hemisphere, around the month of June to September; the northern hemisphere experiences summer at the same time; whereas, when its summer in southern hemisphere, around December to March, the northern hemisphere experiences winter (Figure 81).

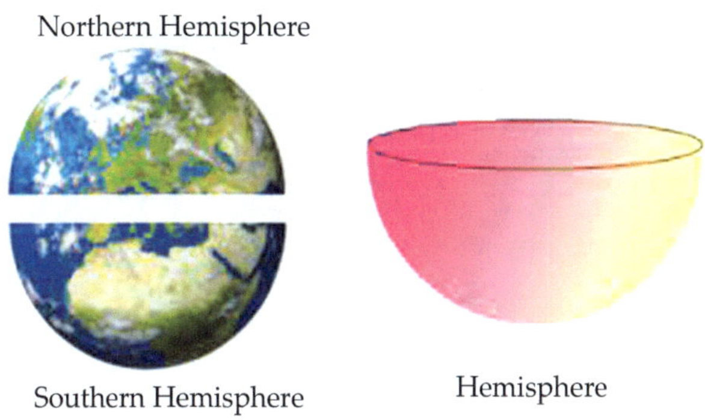

Figure 81. The northern and southern hemispheres.

Source: https://www.google.com/imgres?imgurl=https%3A%2F%2Fi.pinimg.com%2Foriginals%2F1c%2Fd0%2F8c%2F1cd08c3c0c00f5066f4fb6333bc7be10.png&imgrefurl=https%3A%2F%2Fwww.pinterest.com%2Fpin%2F557742735073317926%2F&tbnid=d2dUGWTQ14iRJM&vet=12ahUKEwj6teaK2970AhVGyxQKHealC1QQMygIegUIARDfAQ.i&docid=tYVPVdUrxnfojM&w=456&h=252&itg=1&q=hemisphere&ved=2ahUKEwj6teaK2970AhVGyxQKHealC1QQMygIegUIARDfAQ.

In terms of population, the northern hemisphere has a much higher population compared to the southern hemisphere. The factor of population reflects the level of industrialization in both hemispheres. The level of industrialization of economic growth in the northern hemisphere is so high compared to the southern hemisphere, which is still trying its best to improve its industrial growth rate and increase its economic growth to a better level (Oreskes, 2013).

Storms in the northern hemisphere occur in an anticlockwise manner, whereas storms in the southern hemisphere occur in a clockwise manner. In the northern hemisphere, the Sun rises to its maximum position on the south side size, because, the equator is to the south of the hemisphere. In the southern hemisphere, the Sun rises to its maximum position on the north side, since the equator is to the north of the hemisphere.

The western hemisphere is made up of America and the islands around both the North America and South America continents. The eastern hemisphere is made of Europe, Asia, and Africa (Figure 82).

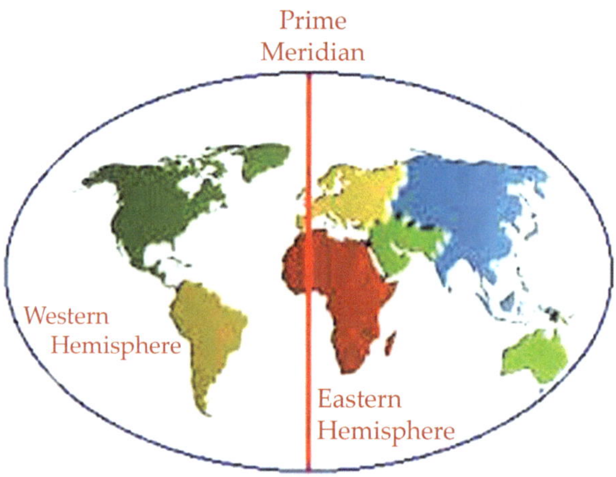

Figure 82. Eastern and western hemisphere.

Source: https://www.google.com/imgres?imgurl=https%3A%2F%2Fqph.fs.quoracdn.net%2Fmain-qimg-d9dee112ad4c225754fd9fb249878962-lq&imgrefurl=https%3A%2F%2Fwww.quora.com%2FWhy-did-the-Eastern-Hemisphere-develop-faster-than-the-Western-Hemisphere&tbnid=y4OmjpEewzI-IM&vet=12ahUKEwir_uur2970AhUPVhoKHcCuDisQMygBegUIARDaAQ.i&docid=epWLfP8sNhM51M&w=362&h=264&itg=1&q=eastern%20and%20western%20hemisphere&ved=2ahUKEwir_uur2970AhUPVhoKHcCuDisQMygBegUIARDaAQ.

High Frequency

Frequency is a one of the terms used to describe the characteristics of a wave. A wave, is used to refer to the variation transfer of energy from one point to another along a medium. The Earth is surrounded by waves, which are referred to as seismic waves. Seismic waves are caused by the movement of the Earth's tectonic plates such as Earthquakes, by explosions, volcanoes, and landslides (Odenwald, 2020). The rate at which the waves generated travel through the Earth's surface could either be categorized as high frequency or low frequency. High frequency is used to describe Earth waves which are being transmitted from the source at a fast rate (Figure 83).

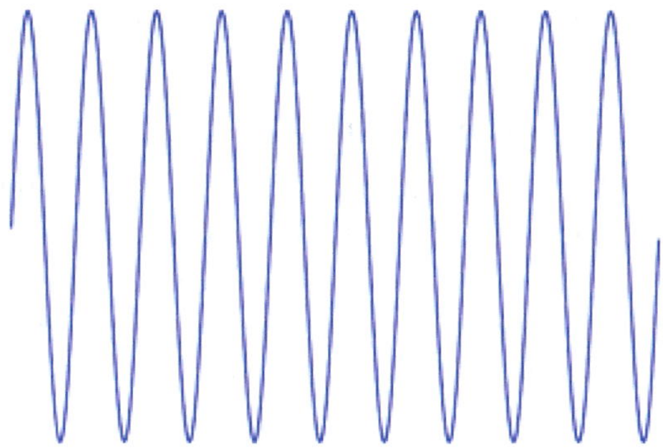

Figure 83. A graph of high frequency.

Source: https://www.google.com/imgres?imgurl=https%3A%2F%2Flabcit.ligo.caltech.edu%2F~jkanner%2Faapt%2Fweb%2Fimages%2Fblue-wave.png&imgrefurl=https%3A%2F%2Flabcit.ligo.caltech.edu%2F~jkanner%2Faapt%2Fweb%2Flesson.html&tbnid=5lHVbNIHPTfdDM&vet=12ahUKEwia0bXe2t70AhVI8IUKHQbFBRUQMygJegUIARDhAQ.i&docid=tHRRIFp-aUPXaM&w=661&h=429&itg=1&q=high%20frequency%20&ved=2ahUKEwia0bX-e2t70AhVI8IUKHQbFBRUQMygJegUIARDhAQ.

Apart from seismic waves, the Earth is also made-up electromagnetic waves, which help in the transfer of light from the Sun to the Earth's surface. The rate of movement of electromagnetic waves through the vacuum of space from the Sun to the Earth is quite high, hence said to have a high frequency. Electromagnetic waves from the Sun to the Earth's surface move as photons. A photon is a light particle moves from its source to a certain destination through electromagnetic radiation (Neeck and Volz, 2011).

The Earth does also allow movement of sound from point source to a destination point. Sound waves vibrating through the air could either have high frequency or low frequency. Sound waves of high frequency are said to be high pitched.

High-Level

High level is a term which could be used in Earth science to refer to a branch which majorly make the Earth science to be a complete category of science. Planet Earth is a complex structure, which is yet to be completely understood; however, the studies done on planet Earth has been categorized into different branches. All the discovered branches so far have been ruled out to be interconnected in one way or the other. The high-level branches of Earth science include: geology, oceanography, climatology, meteorology, and astronomy.

Geology is the study of all the solid materials which make up the Earth, such a rocks, minerals, mountains, and canyons, among others. The science of geology majors on understanding how the different solid materials came into existence; what they are made of; whether they are destructible or indestructible; their interaction with each other; and their importance to human life. Since geology involves the study of several solid materials making up the Earth, it is further divided into different sub-branches, with geologists specializing in the study of each sub-branch; for instance, there are mineralogist who specialize in the study of the different categories of minerals (Movva et al., 2012).

Figure 84. An image of the Earth's geology.

Source: https://www.google.com/imgres?imgurl=https%3A%2F%2Fnews.yale.edu%2Fsites%2Fdefault%2Ffiles%2Fstyles%2Fopengraph_image%2Fpublic%2F125235081-ynews.jpg%3Fitok%3DhodisYP-&imgrefurl=https%3A%2F%2Fnews.yale.edu%2F2020%2F02%2F10%2Fyale-rename-geology-geophysics-department&tbnid=-LHvKeaeEIPghM&vet=12ahUKEwjN7KKy2t70AhW04uAKHVLYBWUQMygAegUIARDQAQ.i&docid=A-qFOEHWdFCyVM&w=766&h=402&q=geologyimages&ved=2ahUKEwjN7KKy2t70AhW04uAKHV-LYBWUQMygAegUIARDQAQ.

Volcanologists specialize in the study of volcanic mountains, what they are made of and how they came into existence. Seismologists specialize in the movement of tectonic plates, which are known to be the main source of seismic waves that cause Earthquakes. Planetary geologists compare different planets in terms of their similarities and differences. Paleontologists specialize in the study of fossils, hoping to understand the different living organisms which ever existed on the Earth. Geology is such a broad level of Earth science (Figure 84).

Oceanography, as the name suggests, is the study of all the components that make up the ocean environment. The Earth is said to be 30% dry ground and 70% water, which is found in large percentages in the oceans. With oceans being the largest percentage of the Earth, so much about them is yet to be discovered. Despite the oceans being so vast, they have immensely been affected by human activities. Overfishing has affected the population of fish found in different oceans.

Figure 85. Different components found in the ocean environment.

Source: https://www.google.com/imgres?imgurl=https%3A%2F%2Fleveragеedu.com%2Fblog%2Fpu%2Fwp-content%2Fuploads%2Fsites%2F3%2F2020%2F03%2FOceanography.jpg&imgrefurl=https%3A%2F%2Fleverageedu.com%2Fblog%2Fpu%2Foceanography%2F&tbnid=q_JuIPntTJlxpM&vet=12ahUKEwiKo7SB2t70AhUSxhoKHaJBDfAQMygEegUIARDUAQ.i&docid=7lq7QslOSyLWdM&w=2560&h=1601&itg=1&q=oceanography%20images&ved=2ahUKEwiKo7SB2t70AhUSxhoKHaJBDfAQMygEegUIAR-DUAQ.

The rate of pollution in oceans has immensely increased, with most parts of the world using the oceans as their dump areas (Müller et al., 2019). General Earth pollution has contributed to global warming, which in turn affects the oceans by melting the ice found in the ice caps; rising the water levels in oceans.

Oceanographers are scientists who specialize in the study of oceans as they look try to understand what happens in the underwater world. Oceanographers are grouped into two different categories. The physical oceanographers specialize in the study of the visible processes which happen in the oceans, such as waves and ocean currents. Chemical oceanographers, as the name suggests, try to understand the chemical components found in the oceans. It is chemical oceanographers who determine the rate of pollution in the oceans (Figure 85).

Climatology, as the name suggests, is the study of the climate. Climatologists are scientists who specialize in the study of the entire atmosphere, helping them understand the climate in general. Atmosphere is the layer of gas surrounding the Earth, and about 11 kilometers above the Earth's surface. The level of thickness of the Earth's atmosphere makes it possible for climatologists to determine temperatures, from the equator outwards; and it contains enough oxygen to support life, unlike in planet Venus where the atmosphere is so thick with high amounts of carbon oxide, making it impossible for the support of life (Metz, 2008). Through the help of climatologists, we are able to understand the level of air pollution taking place on the Earth's surface; the experiments conducted by climatologists identifies greenhouse gases released on the Earth's atmosphere, paving the way for prediction of the possible effects (Figure 86).

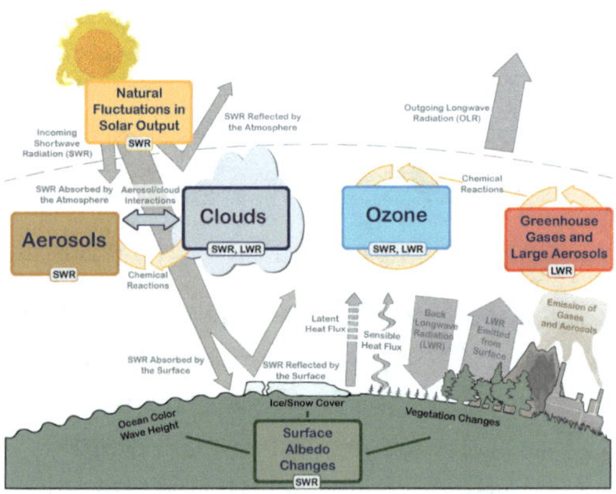

Figure 86. Climate system components.

Source: https://www.google.com/imgres?imgurl=http%3A%2F%2Fwww.climatechange2013.org%2Fimages%2Ffigures%2FWGI_AR5_Fig1-1.jpg&imgrefurl=http%3A%2F%2Fwww.atmos.albany.edu%2Ffacstaff%2Fbrose%2Fclasses%2FATM623_Spring2015%2FNotes%2FLectures%2FLecture04%2520--%2520Climate%2520system%2520components.html&tbnid=HsMxllPvBYp4MM&vet=12ahUKEwiQxJvP

2d70AhVL_hQKHTgoCuAQMygAegUIARCqAQ.i&docid=0mjrT-PU7PY3BM&w=2063&h=1583&itg=1&q=climatology%20components&ved=2ahUKEwiQxJvP2d70AhVL_hQKHTgoCuAQMygAegUIARCqAQ.

Meteorology is similar to climatology for it also deals with the study of atmosphere. However, the main difference between the two is that, meteorology focus on weather patterns, clouds, hurricanes, and tornadoes; whereas climatology focus on study of general climatic conditions, which take a long time to change, mostly a period of 15 to 30 years (Li and Li, 2016). Meteorologists help in predicting the daily weather patterns. They use technological equipment such as radars and satellites which help in predicting weather patterns (Figure 87).

Figure 87. Meteorology components.

Source: https://www.google.com/imgres?imgurl=https%3A%2F%2Fadmin.goweatherforecast.com%2FtemplateEditor%2Fkcfinder%2Fuploads%2Fimages%2F5.2021%2Felements-of-weather-and-instruments-measuring-them.jpg&imgrefurl=https%3A%2F%2Fgoweatherforecast.com%2Fnews%2Felements-of-weather-and-instruments-measuring-them-135&tbnid=0dvWb_zbWuKgjM&vet=12ahUKEwjc6qOz2d70AhVD44UKHeBSApcQMygHegUIARDcAQ.i&docid=hfoPGAfz7OFg8M&w=640&h=427&itg=1&q=meterology%20components&ved=2ahUKEwjc6qOz2d70AhVD44UKHeBSApcQMygHegUIARDcAQ.

Astrology is the science that specializes in the study of the space, all the heavenly objects surrounding the Earth, and in the same surrounding as the Earth. For quite a long time, space was known to be made up of only the solar system. However, with the help of advancing level of technology, astronomers have been able to discover so many planets away from the famous solar system; so many black holes have been discovered; millions of galaxies which were not known; asteroids, comets, and nebula. The space is so vast, with the help of astrology, we are given a chance to stop being obsessed with our planet Earth, and appreciate the diversity and complexity of the universe (Ledley et al., 2011). With the help of telescopes, both optical and radio, we are able to see things in space, which cannot be seen only with human eyes. Space travel has allowed astronomers to explore faraway places, helping them lay their hands on new information with each space travel. With astrology, we are able to understand the effect of space activities on our planet Earth (Figure 88).

Figure 88. An astrology chart.

Source: https://www.google.com/imgres?imgurl=https%3A%2F%2Fimgix.bustle.com%2Fuploads%2Fshutterstock%2F2020%2F4%2F17%2F21122ee5-d50b-4379–8cf6–4e631e9098dc-shutterstock-675656857.jpg%3Fw%3D800%26fit%3Dcrop%26crop%3Dfaces%26auto%3Dformat%252Ccompress&imgrefurl=https%3A%2F%2Fwww.bustle.com%2Fp%2Fhow-to-start-reading-your-own-astrology-chart-when-youre-bored-at-home-22762474&tbnid=c9cUc_tab-ryFM&vet=12ahUKEwiZ_-362N70AhUENRoKHTlcA4MQMygZegUIARCSAg.i&docid=9XV-Wk0XIfDmyM&w=800&h=600&itg=1&q=astrology%20images&ved=2ahUKEwiZ_-362N70AhUENRoKHTlcA4MQMygZegUIARCSAg.

Horizon

Horizon is the term used to refer to the line which separates the Earths from the sky. Horizons could be categorized into two. The first category is the Earth-sky horizon, which separates the Earth from the sky. It is further divided into local horizon, geographic horizon, and sea-level horizon. The local horizon, also referred to as the geometrical horizon, include the visible things which distinctively separate the Earth from the sky. It includes trees, buildings, and mountains. Geographic horizon, is the invisible separation between the Earth and the sky (Lewis, 2008). By invisible, it means that it cannot be seen with human eyes. Sea-level horizon can be considered to be the geographic horizon of the sea, at the sea level. While on the beach, a flat line is visible, which shows where the Earth seems to meet with the sky; and the part of the see which provides this reflection is referred to as an offing (Figure 89).

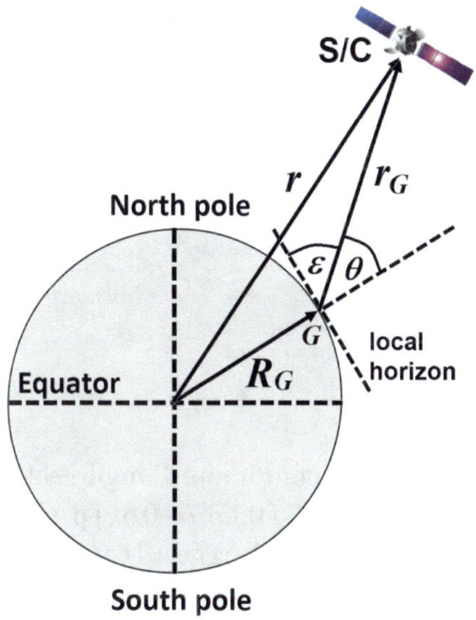

Figure 89. A local horizon.

Source: https://www.google.com/imgres?imgurl=https%3A%2F%2Fwww.researchgate.net%2Fprofile%2FElena-Fantino%2Fpublication%2F339065927%2Ffigure%2Ffig13%2FAS%3A855418624544769%401580959101320%2FDefinition-of-local-horizon-for-a-point-G-on-the-surface-of-Enceladus-and-the.ppm&imgrefurl=https%3A%2F%2Fwww.researchgate.net%2Ffigure%2FDefinition-of-local-horizon-for-a-point-G-on-the-surface-of-Enceladus-and-the_fig13_339065927&tbnid=ntPtpJ9f-iEXCM&vet=12ahUKEwiHvcza2N70AhW_gc4BHeJ4DfEQMygEegUIARC1AQ.i&docid=jgbXnaS8wjI-FM&w=836&h=1068&itg=1&q=local%20horizon&ved=2ahUKEwiHvcza2N70AhW_gc4BHeJ4DfEQMygEegUIAR-C1AQ.

The second category is the celestial horizon, which as the name suggests, is the measurement of the Earth from the sky. It is further divided into astronomical horizon and true horizon. Astronomical horizon is the imaginary separation believed to be a horizontal, and at a 90° angle, directly above the observer (Košler, 2007). The true horizon is also an imaginary separation that is believed to pass through the Earth's surface, being perpendicular to the Earth's radius. Celestial horizons are mostly known to astronomers (Figure 90).

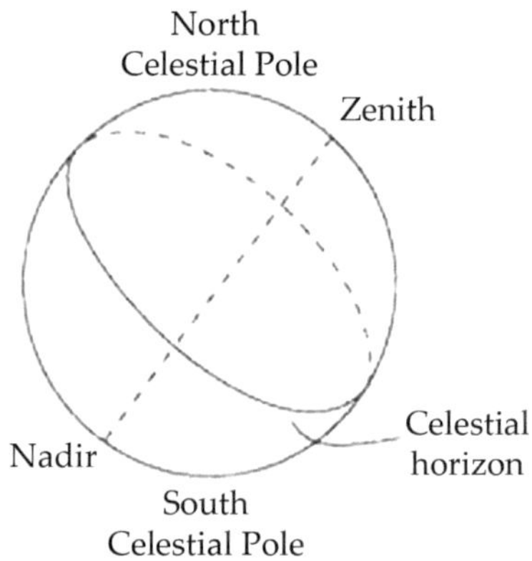

Figure 90. The celestial horizon.

Source: https://www.google.com/imgres?imgurl=https%3A%2F%2Fimg.tfd.com%2Fmgh%2Fav%2Ff0137–05.jpg&imgrefurl=https%3A%2F%2Fencyclopedia2.thefreedictionary.com%2Fcelestial%2Bhorizon&tbnid=UUEUKuK7WYj_WM&vet=12ahUKEwj5xKy42N70AhUHYBoKHdDeD-gQMygBegUIARC-AQ.i&docid=dUm9vioQzHUvkM&w=189&h=195&itg=1&q=celestial%20horizon&ved=2ahUKEwj5xKy42N70AhUHYBoKHdDeD-gQMygBe-gUIARC-AQ.

A horizon, apart from separating the Earth from the sky, are important in the world of aviation, in navigation, and in art. In aviation, pilots use horizons to maintain the plane on a stable position while in the air; by determining the balance of the nose in relation to the horizon; which helps in shifting from one altitude level to another. As for navigations, see navigation before the introduction of the modern navigation tools, relied on the view of horizon, to guide them in navigation. In order to determine time, sea navigators estimated the Sun's position in relation to the horizon (Kleinhans et al., 2010). Celestial horizon was also used by sailors, more so during the night, to help with the navigation for the darkness made it hard to see the sea-level horizon. With the help of celestial bodies such as the stars, the sailors were able to determine celestial horizon. In the art world, artists use the concept of horizon to create an informative painting and drawing; more like creating an illusion.

Humid

Humid, is a term used to refer to high levels of moisture accumulation in the atmosphere. In most cases, when the air is humid, the temperatures are high; because the high temperatures contribute to the evaporation of water from water bodies, and transpiration of water from plants through their stomata; accumulating in the atmosphere (Figure 91).

Figure 91. Humid air.

Source: https://www.google.com/imgres?imgurl=https%3A%2F%2Fmedia.nationalgeographic.org%2Fassets%2Fphotos%2F000%2F216%2F21619.jpg&imgrefurl=https%3A%2F%2Fwww.nationalgeographic.org%2Fencyclopedia%2Fhumidity%2F&tbnid=aEePJ_fwrQK8kM&vet=12ahUKEwi42ZKW2N70AhXq7uAKHUmHDboQMygEegUIARDAAQ.i&docid=PXVPOdTqyrltHM&w=990&h=743&itg=1&q=humid%20air&ved=2ahUKEwi42ZKW2N70AhXq7uAKHUmHDboQMygEe-gUIARDAAQ.

Humidity

Humid is to the act of water accumulation in the atmosphere, whereas humidity is the total amount of water vapor accumulated in the atmosphere. As stated earlier, humidity is determined by temperatures; meaning, a slight change in temperature to the positive side leads to increased humidity, whereas; slight change of temperatures on the negative side, leads to low levels of humidity (Klippel et al., 2019).

Relative humidity is used to refer to the amount of moisture a certain volume air has, compared to the highest amount of moisture it could hold. If the relative humidity is at its highest level, any further increase in temperatures leads to condensation of the moisture into water particles, leading to precipitation. Such a temperature is referred to as dew point. Relative humidity is calculated by dividing the present moisture in the air, with the maximum amount of moisture, the same air volume could hold. The occurrence of fog, mists, and aerosols of water, which in some senses are categorized under humidity, more so relative humidity, are an indication of close to dew point. Relative humidity is always represented in percentages.

Absolute humidity is used to refer to the actual humidity within a certain volume of air at a specific moment. It is measured in grams. It is calculated by dividing the mass of the moisture present in the atmosphere, with the total mass of the air volume available within that region (Hundman and Mattmann, 2017). Just like normal humidity, absolute humidity changes with either change in temperature; or change in pressure is the volume of thc air remains fixed (Figure 92).

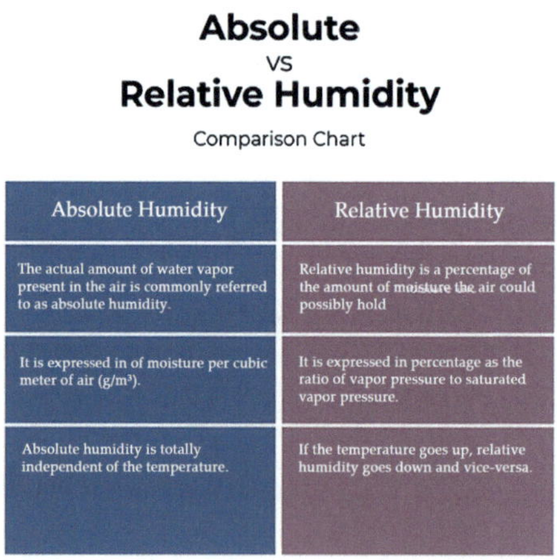

Figure 92. A comparison chart of absolute humidity and relative humidity.

Source: https://www.google.com/imgres?imgurl=http%3A%2F%2Fcdn.differencebetween.net%2Fwp-content%2Fuploads%2F2019%2F05%2FAbsolute-vs-Relative-Humidity.jpg&imgrefurl=http%3A%2F%2Fwww.difference-between.net%2Fscience%2Fnature%2Fdifference-between-absolute-and-relative-humidity%2F&tbnid=oZDPl7v1qmp-mM&vet=12ahUKEwjVlar-S1970AhWGxOAKHRjWDtYQMygJegUIARDcAQ.i&docid=BsYBaJUHv1

0s9M&w=550&h=535&itg=1&q=specific%20humidity&ved=2ahUKEwjVlarS1970AhWGxOAKHRjWDtYQMygJegUIARDcAQ.

Specific humidity, as the name suggests, is the ratio of moisture in the atmosphere to the total moist air parcel. It is also affected by temperature for, as the temperatures move to the negative side, the amount of moisture required to reach saturation level decreases. The vice versa is also true, where, as the temperature moves to the positive side, the saturation level is considered to be close (Figure 93).

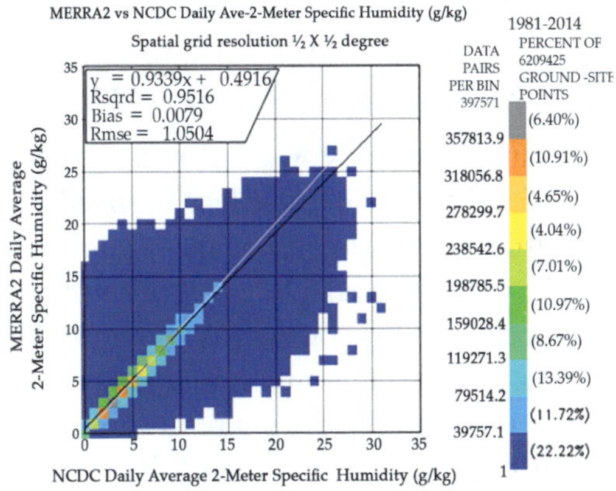

Figure 93. A graph of specific humidity.

Source: https://www.google.com/imgres?imgurl=https%3A%2F%2Fpower.larc.nasa.gov%2Fdocs%2Fmethodology%2Fmeteorology%2Fimages%2FSpecific%2520Humidity.png&imgrefurl=https%3A%2F%2Fpower.larc.nasa.gov%2Fdocs%2Fmethodology%2Fmeteorology%2Fspecific-humidity%2F&tbnid=Ecxu4ZqXEd71ZM&vet=12ahUKEwjVlarS1970AhWGxOAKHRjWDtYQMygGegUIARDWAQ.i&docid=Q885q7HvD4jFAM&w=477&h=373&itg=1&q=specific%20humidity&ved=2ahUKEwjVlarS1970AhWGxOAKHRjWDtYQMygGegUIARDWAQ.

Humidity is measured using a hygrometer. A device known as humidistat could be used in dehumidifiers, which help in lowering humidity in a room to avoid damage to objects which are sensitive to water saturation within their surroundings.

As stated severally in the above section, humidity is affected by temperature; whereby, high temperatures increase humidity, with lower temperatures decreasing humidity. Humidity does also depend on pressure (Guo et al., 2016). High pressure reduces the volume of the air, and in the process, some of the small amount of water aerosols in the atmosphere will condense back into moisture, increasing humidity. Humid air is light in terms of mass compared to dry air. This is because the weight of water molecules in naturally lower, compared to the weight of gases which accumulate dry air.

Humidity does affects other weather factors such as the likes of wind and rainfall. With high humidity, the rate of water saturation will be high, leading to the formation of clouds, which are the sources of rainfall. With low humidity, the chances of rains are quite minimal, unless it is a rainy season. High humidity leads to the occurrence of humid wins, more so winds which are blowing from humid regions, to regions with low humid, carry along the humidity, which in most cases could lead to rains.

Humid air acts as a light lens, for it selectively absorbs the harmful red light from the Sun, only allowing the less harmful green light to pass through it. However, this same property makes humid air be categorized under causes of greenhouse house effects, for it absorbs the infrared energy from the Earth, leading to very low rates of cooling on the Earth's surface, hence high temperatures; an effect common in desert regions, for they only manage to cool during the night, when there are no high temperatures to maintain humidity at a high level. To animal life, high levels of humidity help in reducing the rate of sweating (Goldstein et al., 2017). However, as the humidity increases, the accumulation of warmer blood close the body surface for cooling purposes increases; affecting good circulation of blood within the body, which could lead to a decline in physical strength. Hence, it can be said that humidity plays a factor in classifying which animals are able to survive in which region of the Earth. Same case does also apply to plant life.

The effect of humidity on human beings is similar to that of animals. However, human beings are not that much affected by high humidity levels. Very low humidity causes body tissues such as the likes of the lining of nasal passage to dry, and later crack, making them more susceptible to infection by flu viruses. Extremely low levels of humidity cause eye irritation to human beings. The main advantage of high humidity to human beings is that it reduces the rates of transmission of influenza virus. Exercising in humid regions increases the rate of weight loss to human beings, for it helps in changing the total body water.

In vehicles, high humidity contributes to condensation problems such as shorting of electrical components within the vehicle and misting of windshields. When manufacturing vehicles, and also other vessels such as space ships, a technique of regulating humidity levels ought to be put into considerations (Escorihuela, 2018). In the world of aviation, aircrafts, more long flights, are recommended to have a moderate relative humidity in the cabin space, to avoid effects such as sore eyes, dry skin, and dry nasal tissue. Low humidity, leads to the formation of ice on the aircraft's surface, which could also be a danger, for it affects the total weight of the plane (Figure 94).

Figure 94. Ice accumulation on wingtips due to low humidity.

Source: https://www.google.com/imgres?imgurl=https%3A%2F%2Fflightsafety.org%2Fwp-content%2Fuploads%2F2017%2F02%2FNASA_Flight-tech_feb17_featured_15–097.jpg&imgrefurl=https%3A%2F%2Fflightsafety.org%2Fasw-article%2Fdiscoveries-on-ice%2F&tbnid=jx9zlQwaja58PM&vet=12ahUKEwitzNaR1970AhUItBQKHQezBE8QMygBegUIARCvAQ.i&docid=Nw5Q7hF2jQv6nM&w=824&h=548&itg=1&q=ice%20accumulation%20on%20plane%20wings&ved=2ahUKEwitzNaR1970AhUItBQKHQezBE8QMygBegUIARCvAQ.

I

Iceberg

Iceberg is a block of ice that has severed from glacial mass or shelf ice and is drifting in vast waters. There are much smaller bits of ice called "growlers" or "bergy pieces." Bergy pieces and growlers can start from ice sheets or rack ice, and may likewise be the consequence of an enormous iceberg that has disintegrated. To be called an iceberg, the tallness of the ice should be more than 16 feet above ocean level, and the density should be 98–164 feet, whereas the ice should cover a space of 5,382 square feet.

Figure 95. A massive iceberg drifting in the ocean.

Source: https://www.wired.com/story/an-enormous-iceberg-is-headed-for-south-georgia-island-again/.

There are tinier blocks of ice called "bergy pieces" and "growlers." Bergy pieces and growlers can begin from icy masses or shelf ice, and may likewise be the aftereffect of a huge ice shelf that has separated. The bergy piece is a moderate to huge piece of ice. Its stature is roughly three feet high but 16 feet above ocean level, and its area is roughly around 1,076–3,229 square feet. Growlers are moderate sections of ice and are generally the dimension of a truck or car. They stretch three feet below the ocean surface and cover a space of around 215 square feet (Figure 95) (Bohle, 2015).

Icebergs are by and large white since they are shrouded in snow, yet can be blue, green, yellow, striped, black, and even rainbow-colored. Green growth, seawater, and absence of air bubbles in the ice can make assorted colors. Sediment can make the dark tinge present in some icebergs. Different forms of icebergs exist including: (a) Tabular; (b) Wedge; (c) Pinnacled; (d) Blocky Tabular; (e) Dome; and (f) Drydock ice.

The two primary forms of iceberg structures are even and non-even. Even ice shelves have steep sides and a level top, similar as a plateau, with a length-to-stature proportion of more than 5:1. This sort of ice sheet, otherwise called an ice island, can be very huge, as on account of Pobeda Ice Island (Zhuang et al., 2019). Antarctic chunks of ice shaped by severing from an ice rack, for example, the Filchner-Ronne Ice Shelf, are normally even. The biggest ice shelves on the planet are shaped this way.

Non-even icebergs have various shapes and include: (i) Dome: An ice shelf with a circular top. (ii) Pinnacle: An ice sheet with at least one towers. (iii) Wedge: A chunk of ice with a sharp edge on one side and a slant on the opposite side. (iv) Dry-Dock: An icy mass that has disintegrated to form a space or channel. (v)Blocky: An ice mass with steep, vertical sides and a level top. It contrasts from plain chunks of ice in that its proportions, the proportion between its width and height, is minimal, much like that of a square than a level sheet.

Igneous Rock

Igneous rocks develop when magma (liquid stone) cools and takes shape, either at volcanoes on the outer layer of the Earth or when the dissolved stone is as yet inside the outside. All magma forms underground, in the lower cover or upper mantle, due to the extraordinary temperature there (Yu and Gong, 2012). The two principal classifications of molten rocks are extrusive and intrusive. Extrusive rocks are formed on the outer layer of the Earth from magma risen up out of the ground. Intrusive rocks are formed from magma that cools and settles inside the hull of the Earth. When magma emerges from a spring of gushing lava and cements into extrusive molten stone, also called volcanic, the stone cools rapidly. Precious stones inside strong volcanic rocks are minimal since they do not have a lot of time to form until the stone cools as much as possible, which stops the gem creation. These fine-grained rocks are known as aphanitic—from a Greek word meaning "undetectable." They are given this name on the grounds that the precious stones that form inside them are minor that they can be seen distinctly with a magnifying instrument. Assuming magma cools immediately, the stones that form are smooth with no singular gems, similar to obsidian. There are numerous different categories of extrusive volcanic rocks. For instance, Pele's hair is lengthy, remarkably thin strands of volcanic glass, whereas pahoehoe is smooth magma that develops sparkly, rounded heaps.

Intrusive rocks, commonly called plutonic rocks, cool gradually while never arriving at the surface. They have enormous crystals that are generally perceptible without a magnifying instrument (Witze, 2013). This surface is called a phaneritic surface. Maybe the most popular phaneritic stone is granite. One extreme form of phaneritic rock is known as pegmatite. Pegmatite may have a diverse range of crystal shapes and sizes, some of which are bigger than the human palm.

Imbalance

Earth's environment is determined by a fragile balance between the amount of Sun's radiative energy is captured by the climate and surface, including how much heat infrared radiation Earth discharges to the environment. A positive energy imbalance implies the Earth structure is gaining energy, making the planet warm up. An imbalance happens when you have an excess amount of certain things and excessively less of others. Something imbalanced is wrong or messed up. It is out of equilibrium (Wilkinson et al., 2020). The Earth energy imbalance (EEI), the contrast between the level of energy from the Sun present on Earth and the amount bouncing back to space, fills in as a major measurement to permit mainstream researchers and the general population to determine how well the world reacts to the undertaking of managing climate change.

Impermeable

Impermeable soils and rocks, for example, Earth or shale do not permit water to penetrate, this causes water to run off, diminishing river lag duration and expanding flood risk. Porous rocks permit water to pass through them. Impermeable rocks incorporate shales and unfractured volcanic and metamorphic rocks. Tight shale developments, which are impervious rock and nonporous sandstone and limestone structures and exist (normally) at depths of more than 10,000 feet beneath the surface, likewise contain gas and petrol deposits (Wallace-Wells, 2019). While the practicality of a traditional sandstone repository (containing oil as well as natural gases) is dictated by porosity and penetrability, tight shale developments have almost no porosity and porousness. At times, the oil and gas may be found in a few, detached zones within brief distances of one another, however because of the thickness of the rock development, are blocked off through a similar vertical well (Figure 96).

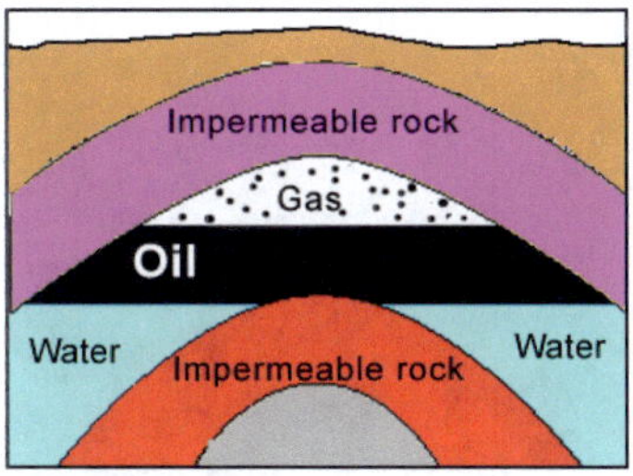

Figure 96. Impermeable rock does not allow water or gas to pass through.

Source: https://www.msnucleus.org/membership/html/jh/earth/petroleum/lesson3/petroleum3a.html.

Inclined

In astronomy, inclination is the point of convergence between two points, one that is an orbital point. The moon circle's plane inclination is 5°9' with regards to the ecliptic plane (the plane of the world's circle around the Sun). The plane's inclination of the ecliptic angle is comparative with the plane of the world's equator at 23°27'8.26"; this point is known as the ecliptic obliquity. The point between the orbital plane of an astronomical body and the reference plane. For comets or planets, the reference plane is the ecliptic plane, for a satellite, it is the tropical plane, and for a twin star, it is the plane of the sky. Inclination is among the orbital components and ranges somewhere in the range of 0 and 180°, being under 90° for a celestial body with direct movement (Toth, Pethe, and Hathazi, 2014).

Inhabit

The meaning of inhabit is to reside in or occupy a spot. One example of inhabit is when raccoons stay in your yard. Inhabited implies any home, or some other structure routinely occupied by creatures, or any external space utilized by individuals consistently. Inhabit is gotten from habitat which is a home space for plants and creatures or different life forms. Examples of environments include: meadow and desert. The primary reasons that motivate animals to inhabit an environment include; food, water, shelter, and space (Figure 97).

Figure 97. Polar bears inhabit the arctic environment.

Source: https://www.google.com/search?q=inhabit+meaning&hl=sw&source=lnms&tbm=isch&sa=X&ved=2ahUKEwjvrOiQ2OD0AhUHhv0HHRheAFoQ_AUoAXoECAEQAw&biw=1366&bih=657&dpr=1#imgrc=NJOfAs1RcQEWwM.

Injection

An injection well is an instrument that places liquid underground into permeable stone structures, like limestone or sandstone, or into or beneath the shallow soil layer. The liquid might be water, wastewater, saline solution (saltwater), or water blended in with synthetic waste (Tollefson, 2015). The liquid that is infused at high depth is occasionally pressurized into faults. When this occurs, liquid pressure rises around the fault, balancing the frictional effects on faults. This may cause earthquakes to happen on them.

An infusion well is utilized to put liquid underground into permeable geologic formations. These underground formations might go from deep sandstone or limestone, to a shallow soil layer. Injected liquids might incorporate water, saline solution (saltwater), wastewater, or water mixed with artificial residues.

Injection well development depends on the kind and depth of the liquid injected. For instance, wells that inject toxic wastes or carbon dioxide (CO_2) into deep secluded structures have complex construction. These wells are intended to give different coats of protective casing and concrete. Alternatively, shallow wells are as a rule of simple makeup.

Intermediate

A volcanic stone with moderate silica portions, and rich in mafic minerals (amphibole, biotite, pyroxene) and felsic minerals (feldspar). Intermediate rocks are dim in shading and contain fairly equal measures of minerals. Intermediate rocks are produced fundamentally in concurrent plate tectonics, in which a maritime plate is subdued underneath either another maritime plate, for example, in Japan or a mainland plate, for example, along the Andean Mountains of South America (Figure 98).

Rocks made generally out of hornblende and intermediate plagioclase feldspars consisting of more than half of all volcanic rocks. Intermediate magmas are fairly stickier than the mafic magmas. Also, they contain fairly a greater amount of gas compared to mafic magmas, yet not exactly as much as the felsic magmas (Thessen, 2016).

Meanwhile, Diorite consists of intrusive intermediate rocks, while, andesite is the name for extrusive intermediate rocks. Intermediate rocks are created basically in focalized plate boundaries in which a maritime plate is subdued underneath either another maritime plate (for example, in Japan) or a mainland plate (for example, along the Andean Mountains of South America).

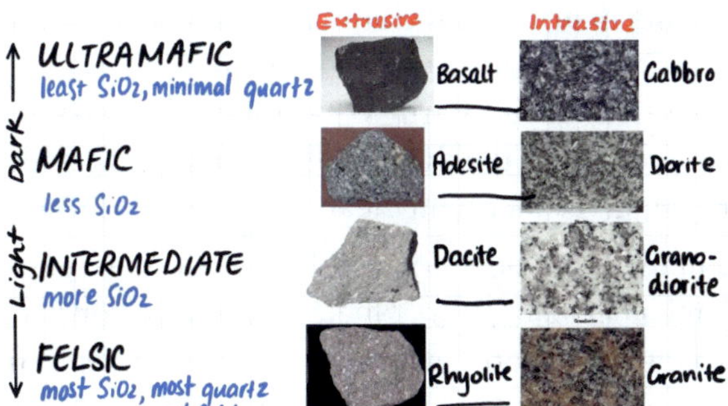

Figure 98. Intermediate soils are a combination of basal and granite.

Source: https://lisbdnet.com/what-type-of-composition-does-an-igneous-rock-have/.

Intermediate magmas will generally form thicker magma streams and higher, more inclined volcanoes (known as stratovolcanoes or composite) as compared to mafic magmas. Examples of composite cones include Mt. Rainier around Seattle, Mt. Fuji in Japan, Washington, and the San Francisco Peaks around Flagstaff, Arizona.

Interplanetary

Being or happening between the planets or between a planet and the Sun. Interplanetary medium is meagerly dispersed matter that exists between the planets and different clusters of the planetary group, just as the natural forces (e.g., magnetic and electric) that spread to an area (Showstack, 2008). The material parts of the interplanetary medium consist of hydrogen, plasma gases containing electrically charged particles from the Sun, vast beams, and residue particles. Incredibly limited quantities of non-ionized hydrogen have been distinguished throughout the interplanetary space.

At the distance of Earth's circle from the Sun, for instance, the centralization of non-ionized hydrogen is around one molecule for every 100 cubic cm (6 cubic inches). A fraction of the neutral hydrogen that enters the nearby planet group from interstellar space is ionized by daylight, and by dispersal of the plasma exuding from the Sun, called the sunlight flare.

The solar flare is a progression of ionized interplanetary gas—particles (primarily protons) and electrons—that constantly grow outward through the adjacent planet group from the Sun's crown. Its depth diminishes with distance from the Sun; at the distance of Earth's circle, it has a thickness of around five particles for every cubic cm (0.06 cubic inch) (Schmitz, 2011). This surge of plasma distributes the attractive fields of power present at the outer layer of the Sun radially away from it. It additionally is responsible for redirecting the tails of the Earth's and other planetary magnetospheres and the tails of comets from the Sun. Those infinite beams identified around Earth involve rapid, high-energy nuclear cores and electrons.

Among the nuclei, the most bountiful are hydrogen (protons; 90%) and helium-based (alpha particles; 9%). Nuclei dwarf electrons around 50 to 1. A minority of astronomical interplanetary beams are delivered from the Sun. A large part of the residue is thought to have been delivered in impacts among space rocks and in the shedding of material from comets while passing close to the Sun (Schmitz, 2012). Around 30,000 tons of interplanetary residue particles are believed to enter Earth's upper atmosphere every year.

Interstellar

An interstellar item is a galactic object (like a space rock, a comet, or a rogue planet, however not a star) within interstellar space that is not gravitationally stuck to a star. This phrase can likewise be applied to an item that is on an interstellar direction, however is briefly passing near a star, like specific space rocks and comets (such as exocomets). In the last option, the item might be known as an interstellar interloper. The original interstellar products were rogue planets (for example, OTS 44 and Cha 110913−773444). The original interstellar object discovered was found within the Solar System, 1I/'Oumuamua in 2017. The next was 2I/Borisov in 2019. The two of them have immense hyperbolic speed, showing they did not begin in the Solar System. Stargazers believe that different interstellar objects of extrasolar beginning (like 'Oumuamua) cross the Earth's orbit each year, and roughly 10,000 passes within Neptune's orbit on any particular day (Rao, 2012). Interstellar comets sometimes cross through the inner Solar System and approach with immense speeds. Until the revelation of 'Oumuamua, no comet with a speed more prominent than the Sun's break velocity had been noticed (Figure 99).

Figure 99. Oumuamua is an interstellar rock that visited the solar system.

Source: https://medium.com/predict/our-first-interstellar-visitor-alien-tech-7106862ee1d6.

Existing models of Oort cloud development foresee that a larger number of comets are shot out into interstellar space than are held in the Oort cloud, with rates of 3 to 100 times as high. Other re-enactments show that above 90% of comets are expelled. There is not a remotely good excuse to believe comets framed in other star systems would not be comparably scattered. The Oort cloud might have been shaped from launched out planetesimals from different stars in the Sun's inception cluster.

It is practical for objects circling a star to be shot out because of connection with a third enormous body, accordingly becoming interstellar objects. Such a cycle was started in the mid-1980s when C/1980 E1, at first gravitationally bound to the Sun, passed close to Jupiter, and was sped up adequately to reach get away speed from the Solar System. This transformed its orbit from circular to hyperbolic and spread out, with a variability of 1.057 setting out toward interstellar space (Rivet and Kastens, 2012). Because of present observational challenges, an interstellar item can typically possibly be identified when it goes through the Solar System, where it very well may be recognized by its unequivocally exaggerated direction and great speed of in excess of several km/s, demonstrating that it is not gravitationally bound to the Sun. Interestingly, gravitationally bound items pursue elliptic orbits across the Sun. (There are a couple of items whose orbits are really near parabolic, so that their gravitationally bound effect is muddled.

Interval

Two structural lines close to each other are divided by a consistent difference in height (like 20 ft and 100 ft). This distinction between contour lines is known as the contour interval. Interface lines connect every one of the map points with a similar height and along these lines uncover the location of slopes, mountains, and valleys (Press, 2008). While a map shows where a street goes, a geographical guide shows why. For instance, the street bends to circumvent a slope or stops at the highest point of a mountain. On an interval map:

- Every interval line addresses a particular height and connects every one of the points that are at a similar elevation. Each fifth interval is bolded and named with mathematical elevations;
- The interval lines run close to one another and do not cross. All things considered; one point can only have a single elevation;
- Two interval lines close to each other are divided by a distinction in height (like 20 ft or 100 ft). This contrast between interval lines is known as the contour interval.

Intrude

An intrusion is a fluid rock that forms under the Earth's surface. Magma from underneath is gradually pushed up from inside the Earth, breaking the rocks above, a process that can require a significant period of time. As the rock gradually cools into a solid, the various pieces of the magma solidify into minerals. Different mountain ranges, such as the Sierra Nevada in California, are shaped generally by intrusion rock. Intrusions are one of the two different ways molten rock can form; the other is extrusive, that is, a volcanic ejection or comparable event valleys (Özyurt et al., 2021). Actually, an intrusion is any arrangement of magma volcanic stone; rock developed from magma that cools and cements inside the external part of the planet. Interestingly, an intrusion comprises of extrusive stone; rock developed from the outer layer of the hull.

Intrusions vary considerably, from mountain-range-scale batholiths to thin veinlike crack fillings of pegmatite or aplite. When uncovered by erosion, these cores known as batholiths might include gigantic expanses of the Earth's surface. Enormous chunks of magma that settle underground before they arrive at the outer layer of the crust are known as plutons. Coarse-grained intrusive molten rocks that develop at the Earth's depth are known as abyssal whereas those that develop close to the surface are known as hypabyssal. Intrusive structures are regularly grouped by whether or not they are corresponding to the sheet material planes, or the country rock foliation: in the event that the intrusion is equal the structure is concordant, in any case it is conflicting.

Invisible

Every one of the stars, planets, and worlds that can be seen today make up only 4% of the universe. The other 96% is made of stuff space experts cannot see, distinguish, or even understand, also called the invisible universe. Astrophysicists additionally concentrate on the invisible universe: at electromagnetic frequencies, more limited than noticeable light; in the gamma beam and at considerably longer frequencies, in the infrared (Orion, 2019). Each wave provides new data. These invisible substances are called dark energy and matter. Space experts surmise their reality dependent on their gravitational effect on what small amounts of the universe can be seen, yet dark matter and energy themselves keep on escaping all discovery (Figure 100).

Figure 100. Dark matter is part of the invisible universe.

Source: https://www.pinterest.es/pin/388928117801176402/.

Dark energy is conceivably considerably more confounding than dark matter. It is a moderately recent discovery, and one that researchers have even to a lesser degree a shot at comprehension at any point in the near future. When we view far off stars, we are seeing an earlier time throughout the entire existence of the universe, on the grounds that those stars' light took millions and billions of light-years to venture out to us. Along these lines, examining the rates of stars at different distances lets people know how quick the universe is growing at different stages in the course of its life.

Ion

An ion is an electrically powered atom or set of atoms developed by the gain or loss of at least one electron. The electron charge is viewed as negative by appearance and this charge is equivalent and contrary to the proton charge, which is viewed as positive by indication. The ion net charge is not zero since its complete number of electrons is inconsistent with its outlier amount of protons.

The cation is a positively charged particle of lesser electrons than protons, whereas an anion is a negatively powered particle of a greater number of electrons than protons. Contrasting electric charges are pulled towards each other by electrostatic power, therefore cations and anions draw in one another and promptly develop ionic mixtures.

Ions comprising of just one iota are called nuclear or monatomic particles, while at least two iotas develop sub-atomic particles or polyatomic particles.

On account of actual ionization in a liquid (gas or fluid), "particle sets" are made by unrestricted atom impacts, where each produced pair comprises of a free electron and a positive ion (Metz, 2008). Ions are additionally made by compound mixtures, like the disintegration of salt in fluids, or by different means, for example, passing an immediate current across a conducting solution, dissolving the anode by means of ionization.

Ions in their gaseous state are exceptionally responsive and will quickly interact with particles of ion charge to provide nonpartisan atoms or ionic salts. Ions are likewise delivered in the fluid or solid-state when salts mix with solvents (for instance, water) to create solvated particles, which are more solid, for reasons including a mix of energy and entropy changes as the particles move from one part to another to bond with the fluid.

As charged particles, ions are drawn to opposite electric charges (positive to negative, as well as the other way around) and repulsed by like charges. When they move, their directions can be determined by an attractive field.

Electrons, because of their more modest weight and hence bigger space-filling properties as waves, decide the size of iotas and atoms that have any electrons whatsoever. In this way, anions (negatively charged particles) are bigger than the parent atom or iota, as the positive electron(s) repulse one another and add to the actual size of the particle, since its size is dictated by its electron cloud (Lewis, 2008). Cations are more modest than the relating parent particle or atom because of the more modest size of the electron cloud. One specific cation (that of hydrogen) contains no electrons, and subsequently comprises of a solitary proton – a lot more modest than the parent hydrogen iota.

Ionized

Whenever weak neural acids and bases are placed in water, they form particles. This is the level of the compound that has ionized (dissociation). Solid acids (bases) ionize completely so their percent ionization is 100%. The percent ionization for bases can be determined (Figure 101) (Clary and Wandersee, 2011).

Most iotas exist in a steady-state where they have an equivalent number of protons and electrons. Notwithstanding, in the event that enough ionization energy is applied, then, at that point, the iota can lose an electron. The measure of ionization energy needed to eliminate an electron fluctuates for each unique component. For instance, chlorine can become ionized by requiring an electron to turn out to be adversely charged. Understanding ionization is vital in concentrating on science since ionization clarifies how the charge of molecules are changed, and can assist with clarifying how particles are altered.

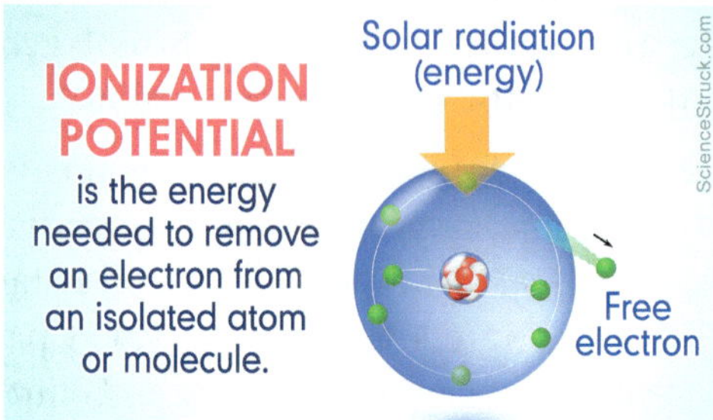

Figure 101. Solar radiation is a powerful ionization agent.

Source: https://sciencestruck.com/ionization-potential.

When HF is put into water, it disintegrates into H+ and F-. This separation can likewise be known as "ionization" as the compound is developing ions. The degree to which this happens is not set in stone from the equilibrium value Ka, yet it can likewise be evaluated by the percent ionization. This is the level of the compound that has ionized (disintegrated). Solid acids (bases) ionize totally so their percent ionization is 100%. Some examples that you might be acquainted to include: sodium and chlorine joining to make salt, the sodium molecule surrenders an electron creating positive charge, while chlorine captures the electron and turns out to be negatively charged (Cossu et al., 2010).

Isolated

An isolated structure is a thermodynamic system that cannot exchange either energy or matter beyond the structure. This system might be so far off from others that it cannot interface with them. The structure might be encased to such an extent that neither energy nor mass might enter or exit. In actual sense, an isolated structure is both of the following:

- An actual structure eliminated from others that it does not mix them;
- A thermodynamic situation through which neither mass nor energy can pass.

An isolated framework is normally taken to be outside the scope of outer gravitational and other long-range powers.

This can be different from what (in the more normal phrasing use in thermodynamics) is known as a closed structure, being encased by specific dividers through which energy can be mistaken for heat, however not make any difference; and with an open structure, which both matter and energy can enter or exit, however it might have differently impermeable dividers in parts of its limits. An isolated system complies with the protection law that its all-out energy–mass remains steady. Frequently, in thermodynamics, mass, and energy are treated as independently preserved.

On account of the necessity of secluded area, and the close to omnipresence of gravity, a rigorous and undeniably disengaged situation do not really happen in tests or in nature.

Isolation

Radiative isolation must be entirely conductive, in order to fully mirror the radiation inside the cavity. In case the cavity with totally reflective dividers contains sufficient radiative power to preserve a temperature of cosmological enormity, then, at that point, the carbon speck is not required in light of the fact that the radiation creates ion particles, such as electron-positron sets, and subsequently arrives at thermodynamic equilibrium (Cerase, 2019).

An alternate methodology is copied by Balian. For quantizing the cavity radiation, he envisions his radiatively detaching dividers to be totally conductive. However, he does not make reference to external mass. As for thermal equilibrium, dividers should contain charged ions that interact with the radiation within the cavity; these cavities are obviously not isolated, yet might be viewed as in a thermal bath.

Isotope

Isotopes are molecules that have a similar number of protons and electrons yet varying quantities of neutrons, and hence have diverse physical properties. Isotopes consist of two types of molecules that have a similar nuclear (amount of protons in their cores) and ranking in the periodic chart (and subsequently have a similar synthetic component), and that contrast in nucleon numbers (mass numbers) because of various quantities of neutrons in their cores. While all isotopes of a given component have practically similar substance properties, they have diverse nuclear masses and physical elements.

The word isotope comes from Greek roots isos meaning (σος "equivalent") and topos (τόπος "place"), indicating "a similar spot;" in this way, the significance behind the name is that various isotopes of a single component possess a similar rank on the periodic table. The quantity of protons inside the iota's core is referred to as atomic number, which is equivalent to the quantity of electrons in the neutral (non-ionized) molecule. Every atomic number distinguishes a particular component, however not the isotope; an iota of a given component may have a wide reach in its number of neutrons (Britner, 2008). The quantity of nucleons (the two protons and neutrons) in the core is the atomic mass number, and every isotope of a particular component has an alternate mass number.

J

Jagged

The word jagged means having a rough edge or surface. Typically, mountains are developed when Earth's crust is shoved up in enormous folds or constrained up or down within blocks. Mountains are jagged since they are continually disintegrated by the climate, which erodes the rough surface away (Butler, 2016). Presumably, most snow-covered mountains are probably going to be decently sharp, a direct result of the disintegration of the water and ice sliding off the side of the mountains; however, and, after its all said and done, assuming the mountain is adequately low to have critical vegetation on its top, then, it very well may have the option to hold together somewhat better.

It will almost certainly likewise depend on the age of the mountain (as the Himalayas are, or alternately are the older mountains that are wearing away). Another element is that regularly when the mountain does not have a sharp pinnacle, it is difficult really to see the top from a good way, as it is frequently darkened by lower sections of the mountain.

It will in general suffice those sharp pinnacles are either unevenly edged like the Alps, or are volcanoes, like Kilimanjaro, while level tops are shaped by rainfall and stream/waterway disintegration based on what was a level scene (Zhuang et al., 2019). Some rough pinnacles can be the aftereffect of ice movement – terminal moraines and drumlins for instance, however these for the most part do not get much over a couple 100 feet high. There are normally exemptions for this. For example, Ingleborough, and Phen-y-Ghent, The Yorkshire Dales and slopes.

Jet Stream

A jet stream of fluid or gas pressured out of a tiny opening. A jet fountain is a rare form of underground aquifer that squeezes and emits, streams of hot water and steam into the air. The bubbling water starts to steam, or go to gas. The steam jets toward the surface. A spring emission is set off when the superheated water fills the fountain's pipe's structure and the spring starts to carry on like a strain cooker. A portion of the water turns to steam. As the steam bubbles become bigger and more copious, they do not really rise openly through the choking influences in the pipes structure (Yu and Gong, 2012).

Springs result from the heating of groundwater by shallow magma. They are by and large connected with areas that have seen past volcanic action. The rambling activity is brought about by the abrupt strain that has been keeping bubbling water in deep, limited conductors underneath a fountain.

Jupiter

Jupiter is the 5th planet from our Sun and is, by a long shot, the biggest planet in the planetary group – over two times as enormous as all the other planets joined. Jupiter's stripes and twirls are really icy, breezy billows of alkali and water, drifting in an air of hydrogen and helium. Jupiter's notorious Great Red Spot is a goliath storm greater than Earth that has seethed for many years (Figure 102).

Figure 102. Jupiter is the largest planet in the universe.

Source: https://www.nasa.gov/audience/forstudents/k-4/stories/nasa-knows/what-is-jupiter-k4.html.

Jupiter is encircled by many moons. Jupiter additionally has a few rings, yet dissimilar to the renowned rings of Saturn, Jupiter's rings are exceptionally weak and made of residue, not frost. Jupiter's current condition is most likely not helpful for life as we know it. The temperatures, pressure, and materials that describe this planet are in all likelihood excessively outrageous and unstable for organic entities to adapt to (Witze, 2013). While planet Jupiter is a far-fetched place for living things to grab hold, the equivalent is not true for a portion of its many moons. Europa is probably the likeliest spot to find life somewhere else in our nearby planet group. There is proof of a huge sea just underneath its cold hull, where life might actually be supported.

With a diameter of 43,440.7 miles (69,911 kilometers), Jupiter is multiple times more expansive than Earth. Assuming Earth were the size of a nickel, Jupiter would be comparably large as a basketball. From an average distance of 484 million miles (778 million kilometers), Jupiter is 5.2 galactic units from the Sun. One galactic unit (defined as AU), is the span from the Sun to Globe. From this distance, it takes Sunlight 43 minutes to venture out from the Sun to Jupiter.

Jupiter has the briefest day in the planetary group. A single day on Jupiter will take around 10 hours (the period it takes for Jupiter to encircle around once), and Jupiter makes a total circle around the Sun (a year in the Jovian period) in around 12 Earth years (4,333 Earth days). It has a tilted equator and its orbital route around the Sun is only 3°. This implies Jupiter rotates almost upright and does not have seasons as intense as different planets do (Wilkinson et al., 2020).

L

Lagoon

A lagoon is a waterway isolated from bigger waterways by a natural boundary; they are separated from bigger waterways by shoals, coral reefs, barrier reefs, or other regular boundaries. Lagoon comes from the Italian word laguna, which means "lake." Although lagoons are distinct topographically, they are in some cases utilized as a name for a bigger location that contains multiple tidal ponds. For instance, Laguna Madre on the Texas Gulf Coast comprises of more modest narrows and tidal ponds, while Laguna Beach in Southern California is really an ocean side and not a lagoon by any stretch of the imagination. There are two kinds of lagoons: atoll and beachfront. Atoll lagoons develop when an island totally submerges under the water, causing a ring of coral that keeps on developing upwards. At the focal point of the ring is a waterway that is often profound. The mix of coral development and water makes a tidal pond. It might take up to 300,000 years for an atoll development to happen. Seaside lagoons form along delicately slanting coasts (Wallace-Wells, 2019). They are for the most part shallower than atoll tidal ponds and will quite often be isolated from the sea by an island, reef, or shoal. More often than not, waterfront lagoons are connected to the sea by a gulf.

Lahar

Lahar is an Indonesian term meaning hot or cold combination of water and rock parts that streams down the inclines of a fountain of liquid magma, and regularly enters a waterway valley. Lahars for the most part occur close to stratovolcanoes, for example, those of the Aleutian volcanic curve in Alaska (Figure 103).

Figure 103. Lahar mudflow on Mount Mayon.

Source: https://theworld.org/stories/2018–01–27/thousands-flee-eruption-heavy-rains-could-create-deadly-mudflows-around-mount.

A moving lahar resembles a slurry of wet cement, and as it surges downstream, the size, speed, and measure of material carried can continually change. The underlying stream might be generally negligible; however, a lahar might fill in volume as it entrains and fuses anything in its way – rocks, soil, vegetation, and even structures and scaffolds. The streaming slurry might devour extra water through liquefying of snow and ice or by inundating waterway or lake water. Voluminous lahars normally develop to in excess of multiple times their underlying size as they move downhill. In steep regions, lahars can surpass rates of 200 km/hr (120 mi/hr), however as they create some distance from a spring of gushing lava and decelerate in swamp regions, they at last start to store a portion of the heap and diminish in size (Toth, Pethe, and Hathazi, 2014).

Land Mass

A landmass is an enormous space of land. The term is regularly used to mean lands encompassed by a sea or ocean, like a landmass or a huge island.

Landmasses incorporate supercontinents, mainland, and islands. There are four significant constant landmasses on Earth: the Americas, Antarctica, Afro-Eurasia, and Australia. In the area of geography, a landmass is a characterized segment of mainland cover stretching above ocean level. Tectonic movements split and separated to form every one of the continents we know today.

Landslide

A landslide is described as the progression of a mass of rock, junk, or Earth down a slant. Landslides are a form of "mass wasting," which signifies any down-slant progression of soil and rock under the immediate impact of gravity. Earthquakes and different elements can likewise prompt landslides. They are brought about by intensifications in the regular steadiness of an incline. They can follow with heavy rains or follow dry seasons, quakes, or volcanic emissions (Bohle, 2015). Landslides are essential for a broader disintegration or surficial procedure of mass wasting, which is just the downslope progression of Earth or surface materials because of gravity. They are subdivided into four fundamental parts: fall and overturning, slides (rotational and translational), streams, and creep.

Lapse Rate

The lapse rate refers to the ratio whereby an atmospheric variable, usually temperature in Earth's environment, falls with elevation. Lapse rate comes from the term lapse, in the feeling of a continuous fall. It compares to the vertical aspect of the spatial slope of temperature. The rate whereby the air temperature shifts with height in the air encompassing a cloud or a rising bundle of air. The average rate is a decline of around 6.5°C/km, however the rate fluctuates significantly in various areas of the world, in different airstreams, and at diverse seasons of the year (Tollefson, 2015). Where the temperature lapse rate is negative (temperature increments with tallness), a reversal is said to exist.

The changing environmental lapse rates all through the Earth's climate are of great significance in meteorology, especially inside the lower atmosphere. They are utilized to decide whether the package of rising air will ascend sufficiently high for its water to consolidate into clouds to shape mists, and, having developed clouds, regardless of whether the air will proceed to rise and form greater shower clouds, and if these clouds will become larger and develop cumulonimbus clouds (thunder mists).

As unsaturated air gets higher, its temperature falls at a dry adiabatic ratio. The dew point likewise drops (because of diminishing pneumatic force) yet considerably more gradually, commonly around −2°C per 1,000 m. Assuming unsaturated air ascends far enough, ultimately its temperature will reach its dew point, and buildup will start to develop. This elevation is called the lifting condensation level (LCL), whereby mechanical lift is available plus the convective condensation level (CCL) when mechanical lift is missing, whereby, the package should be warmed from beneath to its convective heat. The cloud base shall be somewhere around the layer surrounded by these parameters.

The contrast between the dry adiabatic lapse rate and the ratio whereby the dew point drops is roughly 8°C per 1,000 m. Given a distinction in temperature and dew point markings on the ground, one can without much of a stretch determine the LCL by increasing the distinction by 125 m/°C (Thessen, 2016).

Lava

Lava is simply magma that has been ejected from within the Earth onto the surface. Magma might be ejected at a well of lava or through a crack on the rock surface, ashore or undersea, normally at temperatures of between 800 and 1,200°C (1,470 to 2,190°F). The volcanic stone emerging from the ensuing cooling is magma (Figure 104).

Figure 104. Volcanic lava flow.

Source: https://www.collinsdictionary.com/dictionary/english/lava.

A lava stream emits a combination of volcanic debris and different pieces called tephra. In spite of the fact that lava can be up to multiple times thicker than water, with a viscosity like ketchup, lava can stream significant stretches prior to cooling and cementing in light of the fact that lava exposed to air rapidly cools down.

The word lava is Italian and comes from the Latin word labes, which implies a fall or slide. Its initially known use concerns expulsion of magma from beneath the surface was recorded in 1737 by Francesco Serao. He portrayed "a flow of searing lava" as the flow of water and mud down the flanks of the spring of gushing lava (a lahar) following an eruption (Showstack, 2008).

Lava on the Earth's outer layer consists of silicate minerals: typically, feldspars, olivine, pyroxenes, feldspathoids, quartz, amphiboles, and micas. Rare nonsilicate lavas are formed by the dissolving of nonsilicate mineral deposits, and by detachment of magma into immiscible silicate or nonsilicate fluid.

Leach

Leaching fundamentally depicts the course of water carrying dissolvable substances or tiny particles through rock and soil. As water goes through the stone and soil, it mixes with the surface materials. Compounds on the outer layer of minerals may become dissolved. Leaching, in geography, means loss of solvent materials and colloids from the upper soil level by permeating precipitation. The components lost are carried downward (eluviated) and are largely redeposited (illuviated) in a bottom layer. Leaching is the expulsion of dissolvable materials starting with one level in soil then onto the next by means of water progress in the profile (Schmitz, 2011). This is a soil structure unmistakable from the soil shaping sequence of eluviation that is the deficiency of organic and mineral colloids. Leached and elluviated constituents will generally lack topsoil and stored in the subsoil. The soil horizon aggregating leached and eluviated materials is called the illuviation zone. Laterite soil, which forms in areas with high temperature and substantial precipitation, is an illustration of this cycle in real life.

Levee

A levee is a characteristic or counterfeit divider that squares water from going where we do not need it to go. Levees might be utilized to increment accessible land for home or redirect a waterway so the rich soil of a stream or ocean bed might be utilized for horticulture. They keep streams from flooding urban areas in a tempest flood.

A levee is an organic or artificial divider that stops water from going where it should not go. Levees might be utilized to increase accessible land for habitation or redirect a waterway, so the fertile soil of a stream or ocean bed might be utilized for agribusiness. They keep streams from flooding urban areas in a flood. However, assuming that a levee breaks, the results can be shocking.

Levees are typically formed of Earth. The natural course of a waterway, making a traditional levee (Schmitz, 2012). The banks of a waterway are frequently marginally raised from the river bed. The banks develop levees consisting of dregs, residue, and different materials moved aside by the streaming water. Levees are normally corresponding to the manner in which streams flow; therefore, levees can assist with coordinating the progress of the waterway.

Levees can likewise be artificially made or built up. Artificial levees are generally built by heaping soil, sand, and rocks on a cleared, flat surface. In areas where the river flow is solid, levees may likewise be constructed from squares of plastic, metal, and wood. Where the region adjacent to a stream or other waterway is especially risky, levees might even be supported by concrete.

Light Year

Light-year is the duration light travels in one year. Light flashes through interstellar space at 300,000 kilometers per second and 9.46 trillion kilometers a year. It requires 43.2 minutes for daylight to arrive at Jupiter, around 484 million miles away. Light is quick, yet the celestial distances are immense. In 60 minutes, light can travel 671 million miles.

Earth is around eight light minutes from the Sun. The distance at light-speed to the actual edge of our planetary group – the farthest corner of the Oort cloud, a variety of torpid comets away from the furthest planet – would require around 1.87 years. Continue onto Proxima Centauri, our closest neighboring star, and distance goes to 4.25 years at light speed (Rao, 2012).

When commenting about the hugeness of the universe, it is not difficult to throw out huge numbers; however, it is undeniably hard to understand exactly how huge, how far, and how various heavenly bodies truly are.

Our galaxy system consists of gravitationally bound stars, whirling in a twist through space. It is just one of around 2 trillion galaxies in the discernible universe. Groups of them are bound together into superclusters; these superclusters are organized in colossal sheets extending across the universe. Our galaxy contains 100 to 400 billion stars, and is around 100,000 light-years across. That sounds gigantic, and it is, essentially until we begin contrasting it with other worlds. Our adjacent Andromeda galaxy, for instance, is approximately 220,000 light-years wide. Another galaxy, IC 1101, stretches as much as 4 million light-years.

Lithosphere

The lithosphere is the external atmosphere of the Earth. It comprises of a delicate hull and the top layer of the upper mantle. The lithosphere is quite cool compared to other sections of the Earth. It incorporates the fragile upper layer of the mantle as well as the crust, the furthest section of Earth's structure. It is surrounded by the top atmosphere and the asthenosphere (one layer of the upper mantle) beneath (Figure 105).

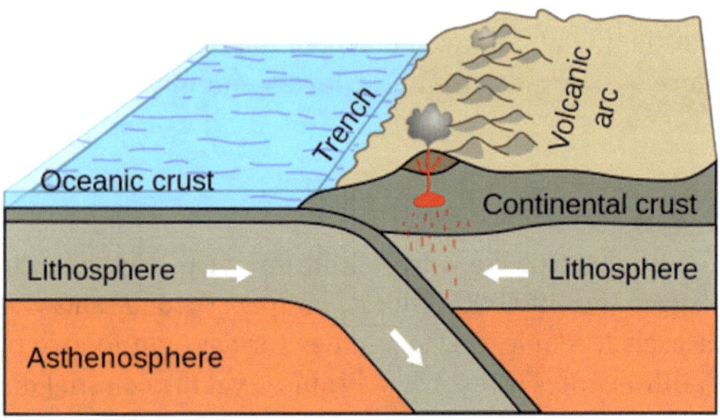

Figure 105. The lithosphere is located directly below the continental crust.

Source: http://www.differencebetween.net/science/differences-between-the-earths-lithosphere-and-asthenosphere/.

While the lithosphere rocks are still flexible, they are not thick. The asthenosphere tends to be vicious, whereas the lithosphere-asthenosphere border (LAB) is the place where geologists and rheologists—researchers who concentrate on the movement of material—mark the distinction in flexibility between the two strata of the upper mantle (Butler, 2016). Pliability measures a strong material's capacity to twist or stretch under pressure. The lithosphere is definitely less flexible compared to the asthenosphere.

There are two kinds of lithosphere: maritime lithosphere and continental lithosphere. Maritime lithosphere is related with maritime oceanic layer, and is marginally denser than mainland lithosphere. The lithosphere likewise connects to the atmosphere, hydrosphere, and cryosphere to impact temperature differences on Earth. Tall mountains, for instance, regularly have significantly lower temperatures than valleys or slopes. The mountain scope of the lithosphere relates with the lower gaseous air pressures and the cold precipitation of the hydrosphere, to establish a cool or even frigid climatic zone.

Low Density

Density is the word for how weighty an item is for its scale. Rock breadth is generally communicated in units such as grams per cubic centimeter (g/cc or g/cm^3), kilograms per cubic meter, and pounds per cubic inch (cubic foot and cubic yard). Rocks differ extensively in density; thus, rock density is regularly a recognizable proof apparatus and helpful for recognizing earthbound (Earth) rocks from shooting stars. Most meteorites are exceptionally thick, 7–8 g/cm^3. These are standard chondrites, and traditional chondrites have a thickness half as much as irons. Most standard chondrites are in the scope of 3.0 to 3.7 g/cm^3, which is denser than most land rocks (Rivet and Kastens, 2012). For instance, limestone (2.6 g/cm^3 or less), quartzite (2.7 g/cm^3), and stone (2.7–2.8 g/cm^3) are typically low-dense rocks. A few meteorites have low densities ($<$3.0 g/cm^3); however, such meteorites are uncommon. The breadth of basalt, quite possibly the most widely recognized form of volcanic rock, can be pretty much as high as 3.0 g/cm^3. Metals like iron, zinc, and lead are also common in meteorites. For instance, rocks made out of (iron oxides) or hematite are frequently confused with meteorites. These rocks have high densities, 4.5–5 g/cm^3, which is more significant than rocky meteorites.

Luminous

Luminous in a real sense means to gleam. Brilliant signs rely upon a gas like neon, krypton, argon, xenon, or radon for luminosity—and you can also utilize iridescent (DayGlo) paint to make your own signs. New innovations have now given us luminous textures, which are being utilized to deliver striking or luminous results in attire, interior surfaces, and upholstery (Figure 106).

Figure 106. A luminous diamond on display.

Source: https://news.centurionjewelry.com/articles/detail/diamonds-alrosa-gears-up-for-new-luminous-diamonds-category-launch-this-fall-remembering-arun-mehta-of-rosy-blue.

Luminous is the unconstrained discharge of light by a substance not coming from heat, as with the case of brilliance, which is light transmitted by a substance because of warming. Luminous is brought about by the assimilation of energy that is delivered in modest quantities. When the energy comes from light or other electromagnetic radiation, it is alluded to as photoluminescence; which is detachable between fluorescence when the shine stops quickly with the excitation and brightness when the gleam proceeds past the period of excitation (Britner, 2008). Two kinds of radiant features are seen in glass-like materials. Triboluminescence produces light through the breaking of compound bonds in a material when it is scoured, pulled, separated, scratched, or squashed. Thermoluminescence re-discharges recently retained electromagnetic radiation after being warmed (e.g., thermoluminescence dating).

Some fluorite, especially the diverse chlorophane (otherwise known as pyroemerald and cobra stone), might become faintly radiant essentially from the temperature of one's hand. Chlorophane is unique for consolidating the properties of thermoluminescence, triboluminescence, and fluorescence; it will transmit visible light when scoured, or presented to light or warmth, and can keep transmitting for an extensive duration of time.

Lunar Eclipse

Lunar eclipses have been a common event for skywatchers for incalculable years. They happen when Earth's shadow impedes the daylight, which in any case bounces off the moon. There are three kinds of lunar eclipses — aggregate, temporary, and penumbral — with the most expressive being an all-out lunar eclipse, whereby Earth's shadow totally covers the moon. During a complete lunar eclipse, the moon will become radiant red during the night sky. Since time immemorial, eclipses have motivated awe and even dread, particularly when absolute lunar eclipses turned the moon crimson, an impact that alarmed individuals who had no comprehension of what causes the overshadowing and subsequently pinned the events onto imaginary gods.

NASA keeps a list of lunar eclipses until 2100. They likewise keep information about past lunar eclipses. During the 21st century, Earth will encounter a sum of 228 lunar shrouds, as indicated by the space organization. The main lunar eclipse of this current era will happen on May 5–6, 2023. This will be a penumbral lunar eclipse which happens when the Sun, Earth, and moon are not well adjusted (Press, 2008). At the point when this occurs, the Earth blocks only a fraction of the daylight from reaching the moon, therefore only the external shadow of the Earth, known as the obscuration, conceals the moon's face. This obscuration will be noticeable, undoubtedly somewhat, from southern and eastern Europe, Antarctica, a large part of Asia, Australia, Africa, just as the Pacific, Indian, and Atlantic oceans. In any case, even where noticeable this sort of obscuration can be hard to notice since it is unobtrusive.

M

Magma

Magma is an extremely hot liquid and semi-liquid rock found beneath the Earth's surface. The Earth's structure is stratified, with an inner core, an outer core, a mantle, and a crust. Magma makes up a large portion of the planct's mantle (Figure 107) (Zhuang et al., 2019).

Figure 107. Lava (magma that has erupted onto the Earth's surface) is visually mesmerizing – as the molten rock flows downhill, lava exposed to the air cools to a deep black color, while the molten rock beneath glows bright orange.

Source: https://www.nationalgeographic.org/encyclopedia/magma/.

This lava has the potential to push through holes or cracks in the crust, resulting in a volcanic eruption. Lava is formed when magma flows or erupts onto the Earth's surface.

Magma, like solid rock, is a mineral combination. Small amounts of dissolved gases such as water vapor, carbon dioxide (CO_2), and sulfur are also present. The extreme heat and pressures beneath the Earth's crust keep magma fluid.

Magma is classified into three types: basaltic, andesitic, and rhyolitic, each with its own mineral makeup. All varieties of magma contain a large amount of silicon dioxide. Iron, magnesium, and calcium are abundant in basaltic magma, whereas potassium and sodium are scarce.

It has a temperature range of roughly 1,000°C to 1,200°C. (1,832°F to 2,192°F). With temperatures ranging from 800°C to 1,000°C, andesitic magma contains moderate levels of these minerals (1,472°F to 1,832°F). Rhyolitic magma has a high potassium and sodium content but a low iron, magnesium, and calcium content. It happens at temperatures ranging from 650°C to 800°C (1,202°F to 1,472°F). Magma's temperature and mineral concentration both influences how easily it flows (Yu and Gong, 2012).

The viscosity (thickness) of the lava that erupts from a volcano influences its shape. Volcanoes with steep slopes are formed by highly viscous magma, whereas flatter volcanoes are formed by magma that flows freely.

Magnesium

Magnesium (Mg) is a chemical element that is the lightest structural metal and one of the alkaline-earth metals of Periodic Group 2 (IIa). Its compounds are widely employed in engineering and medicine, and magnesium is a vital mineral for all cellular life.

The silvery-white element, first known through compounds such as Epsom salts (the sulfate), magnesia or magnesia alba (the oxide), and magnesite (the carbonate), does not occur free in nature. Sir Humphry Davy isolated it for the first time in 1808 when he evaporated the mercury from a magnesium amalgam created by electrolyzing a mixture of moist magnesia and mercuric oxide. Magnesia, a province of Thessaly (Greece) where the mineral magnesia alba was discovered, inspired the term magnesium.

Magnesium is the eighth-most prevalent element in the Earth's crust (approximately 2.5%) and the third most abundant structural metal after aluminum and iron. Its cosmic abundance is calculated to be 9.1 105 atoms (on a scale where silicon abundance = 106 atoms). It can be found as carbonates (magnesite, $MgCO_3$, and dolomite, $CaMg(CO_3)_2$), as well as in several common silicates, such as talc, olivine, and most types of asbestos (Witze, 2013). It is also found in the forms of hydroxide (brucite), chloride (carnallite, $KMgCl_36H_2O$), and sulfate (kieserite). It can be found in minerals including serpentine, chrysolite, and meerschaum. Seawater contains about 0.13% magnesium, largely as dissolved chloride, which contributes to the bitter flavor.

Magnesium is economically generated through the electrolysis of molten magnesium chloride ($MgCl_2$), primarily from seawater, and through the direct reduction of its compounds with suitable reducing agents, such as the reaction of magnesium oxide or calcined dolomite with ferrosilicon.

Magnetic

Magnets are objects with a north pole and a south pole at opposing ends. Electrons in a magnet have both irregular orbits and unequal spins. Inside each domain, the magnetic atoms are neatly organized in straight rows. And those domains are all pointing in the same direction. And it is only when All of these requirements are met that this piece of metal becomes a magnet or magnetic (Wilkinson et al., 2020).

Materials that can be magnetized, which are also the ones that are strongly attracted to a magnet, are called ferromagnetic (or ferrimagnetic). These include the elements iron, nickel, and cobalt and their alloys, some alloys of rare-earth metals, and some naturally occurring minerals such as lodestone.

Magnetic Field

The geomagnetic field, commonly known as the Earth's magnetic field, is the magnetic field that extends from the Earth's interior into space, where it interacts with the solar wind, a stream of charged particles emitted by the Sun. The magnetic field is created by electric currents caused by the movements of convection currents of molten iron and nickel in Earth's outer core: these convection currents are triggered by heat escaping from the core, which is a natural phenomenon known as a geodynamo. At the surface, the magnitude of the Earth's magnetic field ranges from 25 to 65 T. (0.25 to 0.65 gauss). It is approximated by a magnetic dipole field currently tilted at an angle of around 11° with regard to Earth's rotational axis, as if a huge bar magnet were put at that angle through the center of (Wallace-Wells, 2019).

The North geomagnetic pole actually represents the South pole of Earth's magnetic field, and vice versa (because opposite magnetic poles attract and the north end of a magnet, like a compass needle, points toward Earth's South magnetic field, i.e., the North geomagnetic pole near the Geographic North Pole). The North geomagnetic pole was located on Ellesmere Island, Nunavut, Canada, in 2015.

While the North and South magnetic poles are normally near the geographic poles, they shift slowly and constantly throughout geological time spans, but slowly enough for common compasses to remain effective for navigation. However, at periodic intervals of several 100,000 years, the Earth's magnetic field reverses and the North and South Magnetic Poles abruptly exchange locations. These geomagnetic pole reversals leave a record in rocks that paleomagnetists can use to calculate geomagnetic fields in the past. This information is therefore useful in researching the movements of continents and ocean floors during the plate tectonics process.

The magnetosphere is the region of space above the ionosphere determined by the extent of the Earth's magnetic field. It stretches into space for tens of thousands of kilometers, shielding Earth from charged particles in the solar wind and cosmic rays that would otherwise strip away the upper atmosphere, including the ozone layer that protects Earth from the harmful ultraviolet (UV) radiation (Figure 108).

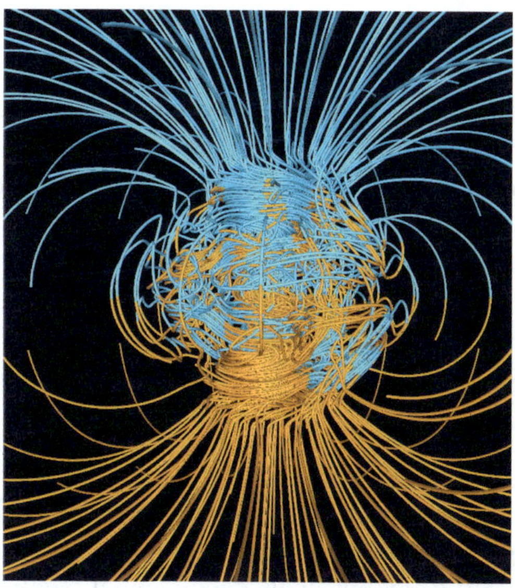

Figure 108. Computer simulation of Earth's field in a period of normal polarity between reversals. The lines represent magnetic field lines, blue when the field

points towards the center and yellow when away. The rotation axis of the Earth is centered and vertical. The dense clusters of lines are within the Earth's core.

Source: https://en.wikipedia.org/wiki/Earth's_magnetic_field.

Magnetism

Magnetism is a set of physical properties mediated by magnetic fields. Electric currents and elementary particle magnetic moments combine to form a magnetic field, which operates on other currents and magnetic moments. Magnetism is a component of the larger phenomenon of electromagnetic Earth (Toth et al., 2014). The most well-known effects occur in ferromagnetic materials, which are highly attracted by magnetic fields and can be magnetized to become permanent magnets, hence producing magnetic fields. It is also feasible to demagnetize a magnet. Only a few materials are ferromagnetic; the most frequent are iron, cobalt, nickel, and their alloys. Less common examples include the rare-earth elements neodymium and samarium.

Magnitude

Magnitude is a fundamental concept in science. The term magnitude refers to the overall quantity or distance. When it comes to movement, we can associate the magnitude, size, and speed of an object while it is in motion. The magnitude of an object is defined by the size or amount of that object.

The magnitude of an earthquake is a number that describes its relative size. Magnitude is determined by measuring the maximum motion recorded by a seismograph. In astronomy, the brightness of a star or other celestial body as seen with the unassisted eye is expressed by a mathematical ratio of 2.512: a star of the first magnitude is approximately 212 times brighter than one of the second magnitude and 100 times brighter than one of the sixth magnitude (Tollefson, 2015). With the naked eye, only stars of the sixth magnitude or brighter may be seen.

Marine

The study of the history and structure of the ocean floor is known as marine geology or geological oceanography. It entails research of the ocean floor and coastal zone using geophysical, geochemical, sedimentological, and paleontological methods. Marine geology is closely related to geophysics and physical oceanography (Figure 109) (Thessen, 2016).

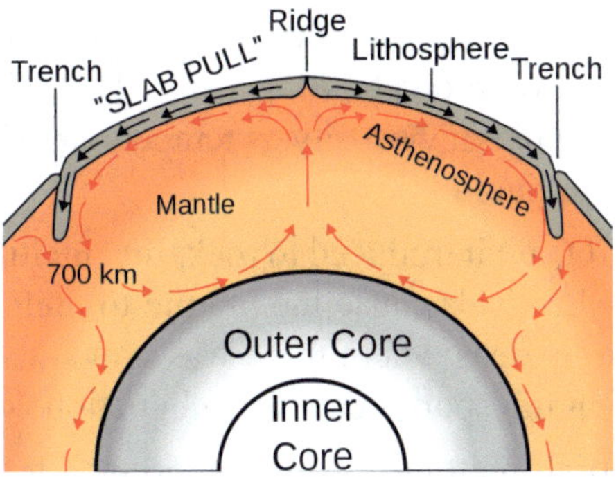

Figure 109. Oceanic crust is formed at an oceanic ridge, while the lithosphere is subducted back into the asthenosphere at trenches.

Source: https://en.wikipedia.org/wiki/Marine_geology#/media/File:Oceanic_spreading.svg.

In the years following World War II, marine geological research was essential in giving critical evidence for seafloor spreading and plate tectonics. The deep ocean floor is the final fundamentally unknown frontier, and thorough mapping is driving the study in support of both military (submarine) and economic (petroleum and metal mining) aims.

Mechanical

Mechanical weathering refers to mechanical processes that break up a rock, such as ice freezing and expanding in cracks in the rock, tree roots growing in similar cracks, rock expansion and contraction in areas with high daytime and low nighttime temperatures, rock cracking in forest fires, and so on.

Mechanical weathering reduces the size of rocks without altering their content. Ice wedging and abrasion are two key mechanical weathering mechanisms. Chemical weathering degrades rocks by creating new minerals that are stable on Earth's surface.

The strength or stiffness of the Earth's mechanical layers distinguishes them. These layers are not the same as the compositional layers of the Earth, such as the crust, mantle, and core, though they share some boundaries.

Melt

Melting is the transformation of a solid into a liquid caused by the application of heat. This process occurs at a set temperature known as the melting point in a pure crystalline solid; an impure solid generally melts over a range of temperatures below the melting point of the primary component (Telling et al., 2017). Amorphous (non-crystalline) materials, such as glass or pitch, melt by gradually reducing in viscosity as temperature rises, with no abrupt transition from solid to liquid.

When water or CO_2 are introduced to rock, flux melting occurs. At lower temperatures, these chemicals cause the granite to melt. This results in the formation of magma in areas where there was previously a solid structure. Flux melting, like heat transport, occurs at subduction zones.

In geology, partial melting is an important step in the chemical differentiation of crustal rocks. On Earth, partial melting of the mantle at mid-ocean ridges (MORBs) results in oceanic crust, while partial melting of the mantle and oceanic crust at subduction zones results in continental crust.

Meltwater

Meltwater is the water that is discharged when snow or ice melts, including glacial ice, tabular icebergs, and ice shelves over oceans. Meltwater is frequently observed in glacier ablation zones, where the rate of snow cover is decreasing. Meltwater can occur during volcanic eruptions in the same way as more severe lahars do.

Melt ponds arise when meltwater pools on the surface rather than flowing. Meltwater frequently re-freezes as the weather cools. Meltwater can gather or melt beneath the surface of the ice. These water pools, known as subglacial lakes, can arise as a result of geothermal heat and friction (Showstack, 2008).

Meltwater provides drinking water for a large proportion of the world's population, as well as providing water for irrigation and hydroelectric plants. This meltwater can originate from seasonal snowfall, or from the melting of more permanent glaciers. Climate change threatens the precipitation of snow and the shrinking volume of glaciers.

Mercury

Mercury is the smallest and nearest planet to the Sun in the Solar System. Its orbit around the Sun is the shortest of all the Sun's planets, taking 87.97 Earth days. It is named after the Roman god Mercurius (Mercury), who was a god of commerce, a messenger of the gods, and a mediator between gods and mankind, similar to the Greek god Hermes. Mercury, like Venus, orbits the Sun as an inferior planet within Earth's orbit, and its apparent distance from the Sun as seen from Earth never surpasses 28°. Because the planet is so close to the Sun, it can only be seen at the western horizon after sunset or the eastern horizon before sunrise, usually in twilight. It may seem as a bright star-like object at this time, but it is more difficult to observe than Venus.

Mercury is also a chemical element with the atomic number 80 and the symbol Hg. It was once known as hydrargyrum and was generally known as quicksilver. Mercury, a heavy, silvery d-block element, is the only metallic element that is liquid under standard temperature and pressure conditions; the only other element that is liquid under these conditions is the halogen bromine, though metals such as Caesium, gallium, and rubidium melt just above room temperature (Schmitz, 2011).Mercury can be found in deposits all throughout the world, primarily as cinnabar (mercuric sulfide). Vermilion is created by grinding natural cinnabar or synthetic mercuric sulfide (Figure 110).

Figure 110. This is a cropped bottom right image from the original four image mosaic PIA11364: Mercury's "True" Color is in the Eye of the Beholder.

Source: https://en.wikipedia.org/wiki/Mercury_(planet)#/media/File:Mercury_in_true_color.jpg.

Mesa

A mesa is a mountain or hill with a flat top. It is a broad, high landform with steep sides. Mesa is a Spanish term for table. The term was coined by Spanish explorers of the American southwest, where many mesas may be found, since the tops of mesas resemble the tops of tables (Figure 111).

Figure 111. The "Great White Throne" is a rock formation in Zion National Park, Utah. The Great White Throne is a mesa, or flat-topped mountain, made of limestone. Zion Canyon was carved over 13 million years by the flow of the Virgin River.

Source: https://www.nationalgeographic.org/encyclopedia/mesa/#:~:text=A%20mesa%20is%20a%20flat%2Dtopped%20mountain%20or%20hill.,Spanish%20word%20that%20means%20table.&text=Mesas%20are%20formed%20by%20erosion,the%20top%20of%20a%20hill.

Mesas are generated through erosion, which occurs when water washes smaller, softer rocks away from the top of a hill. Caprock is the hard, long-lasting rock that remains on top of a mesa. Mesas are typically wider than they are tall (Schmitz, 2012).

Mesas are typically found in dry areas with horizontal rock layers. The Grand Mesa in Colorado, considered the world's largest mesa, covers an area of about 1,300 square kilometers (500 square miles) and runs for 64 kilometers (40 miles).

Mesosphere

The mesosphere is the region of the upper atmosphere between 50 and 80 kilometers (30 and 50 miles) above the Earth's surface. The temperature maximum at the top of the stratosphere is described as the base of the mesosphere, with the boundary between the two zones commonly referred to as the stratopause. The mesosphere extends upward to the next temperature low, which forms the base of the thermosphere; the mesopause is the boundary between the two zones.

The percentages of oxygen, nitrogen, and CO_2 in the mesosphere's air are nearly identical to those in the levels of the Earth's atmosphere directly above the Earth's surface. The main distinctions are that the air density is substantially lower in the mesosphere, there is very little water vapor in the mesosphere, and the mesosphere has larger percentages of ozone than the lower levels (Figure 112) (Rao, 2012).

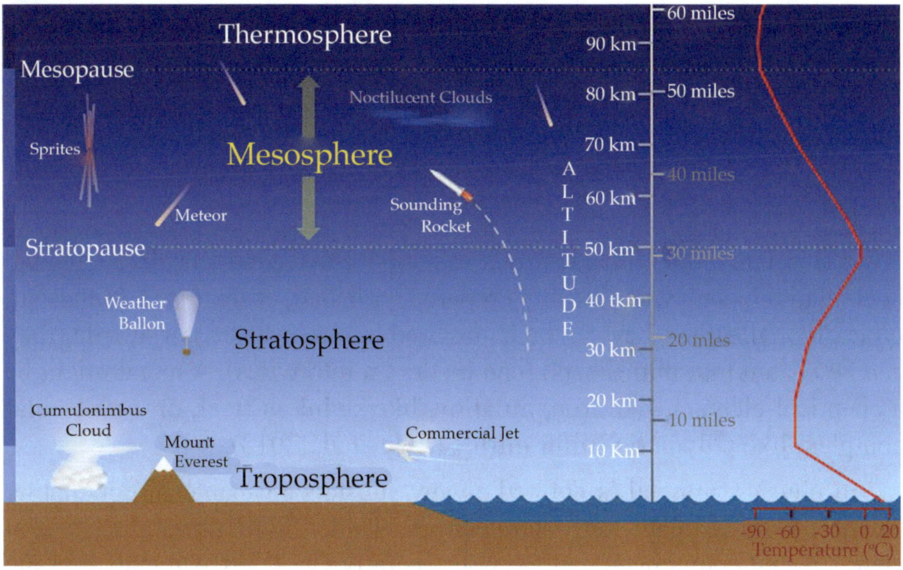

Figure 112. Mesosphere diagram.

Source: https://scied.ucar.edu/image/mesosphere-diagram.

Mesozoic

The Mesozoic Era was the second of Earth's three major geologic eras during the Phanerozoic period. It gets its name from the Greek word for "middle life." The Mesozoic Era began 252.2 million years ago, after the Paleozoic Era ended, and ended 66 million years ago, with the start of the Cenozoic Era. The Mesozoic Era is divided into three primary periods: the Triassic Period, the Jurassic Period, and the Cretaceous Period. The ancestors of today's major plant and animal groups initially arose during the Mesozoic, but this period is best remembered as the time of the dinosaurs (Rivet and Kastens, 2012).

The Earth's climate was usually warm during the Mesozoic Era, with less temperature variance between equatorial and polar latitudes than there is today. The Mesozoic epoch was a period of geologic and biological transformation.

The continents began to shift into their current configurations during this time period. A unique modernization of life-forms occurred, owing in part to the extinction of many previous sorts of organisms. The Mesozoic is associated with three of the five largest mass extinctions in Earth history: one occurred at the boundary between the Mesozoic and the preceding Paleozoic; another occurred within the Mesozoic at the end of the Triassic Period; and a third occurred at the boundary between the Mesozoic and the subsequent Cenozoic, resulting in the extinction of the dinosaurs.

Metal

A metal is a material with a lustrous appearance that conducts electricity and heat relatively well when freshly prepared, polished, or fractured. Metals are typically malleable (they can be hammered into thin sheets) or ductile (they can be beaten into thin sheets) (can be drawn into wires). A metal might be a chemical element, like iron, an alloy, like stainless steel, or a molecular complex, like polymeric sulfur nitride (Pyle et al., 2018).

In physics, a metal is defined as any substance capable of conducting electricity at absolute zero temperature. Under high pressures, several elements and compounds that are not generally classified as metals become metallic. At pressures ranging from 40 to 170,000 times atmospheric pressure, nonmetal iodine, for example, progressively transforms into a metal. Similarly, some metals can transform into nonmetals. At slightly under 2 million times atmospheric pressure, sodium, for example, becomes a nonmetal.

Due to their chemistry, two elements that would ordinarily qualify as brittle metals (in physics)—arsenic and antimony—are usually referred to as metalloids in chemistry (predominantly non-metallic for arsenic, and balanced between metallicity and non-metallicity for antimony). Metals account for approximately 95 of the 118 elements in the periodic table (or are likely to be such). The figure is imprecise because the boundaries between metals, nonmetals, and metalloids vary slightly due to a lack of globally accepted definitions for the groups involved (Press, 2008).

In astrophysics, the phrase "metal" refers to all chemical components in a star that are heavier than helium, rather than just typical metals. In this sense, the first four "metals" to accumulate in star cores via nucleosynthesis are carbon, nitrogen, oxygen, and neon, all of which are chemically nonmetals. During its lifetime, a star fuses lighter elements, usually hydrogen and helium, into heavier atoms. Used in that sense, the metallicity of an astronomical object is the proportion of its matter made up of the heavier chemical elements.

Metals, as chemical elements, comprise 25% of the Earth's crust and are present in many aspects of modern life.

Metamorphic

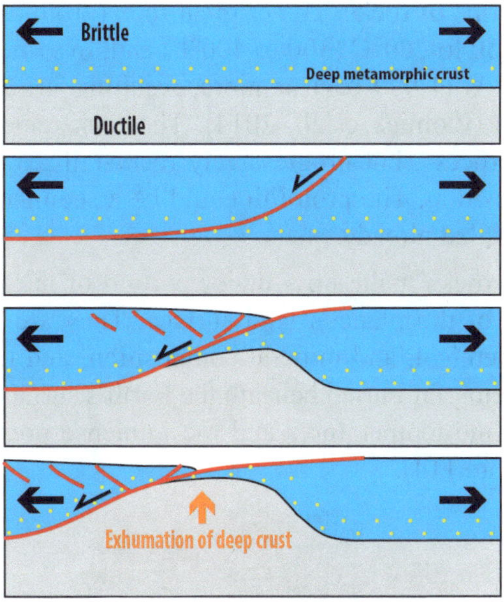

Figure 113. Formation of a metamorphic core complex.

Source: https://en.wikipedia.org/wiki/Metamorphic_core_complex#/media/File:Core_complex.png.

The term metamorphic can refer to a variety of things:

- A metamorphic zone is a geological area where the same mineral combination occurs in the bedrock as a result of metamorphism. Because most metamorphic minerals are only stable at particular temperatures and pressures, these zones form.
- Metamorphic facies are a group of mineral assemblages generated in metamorphic rocks under similar pressures and temperatures. The assemblage is typical of what occurs in conditions corresponding to a region on the two-dimensional temperature

vs. pressure graph. Rocks containing specific minerals can thus be related to specific tectonic settings, times, and locations in the area's geological history. Because the borders between facies (and related areas on the temperature vs. pressure graph) are gradational and imprecise, they are broad (Figure 113).

Metamorphic Rock

Metamorphic rocks form as a result of the metamorphism of existing rocks into new forms of rocks. The original rock (protolith) is subjected to temperatures of 150 to 200°C (300 to 400°F) and, in some cases, pressures of 100 megapascals (1,000 bar) or more, resulting in severe physical or chemical changes (Plotnick et al., 2014). The rock remains mainly solid throughout this process, but progressively recrystallizes to a new texture or mineral composition. The protolith could be a sedimentary, igneous, or metamorphic rock that already exists.

Metamorphic rocks make up a major portion of the Earth's crust and account for 12% of the planet's land surface. They are distinguished by their protolith, chemical, and mineral composition, and texture. They can form simply by being far buried beneath the Earth's surface, where they are subjected to extreme temperatures and the immense pressure of the rock layers above (Figure 114).

Figure 114. Quartzite, a type of metamorphic rock.

Source: https://en.wikipedia.org/wiki/Metamorphic_rock#/media/File:Quartzite.jpg.

They can also occur as a result of tectonic events such as continental collisions, which generate horizontal pressure, friction, and distortion. Metamorphic rock can form locally when rock is heated by the intrusion of hot molten rock from the Earth's interior, known as magma. The study of metamorphic rocks (which have been exposed at the Earth's surface as a result of erosion and uplift) gives information on the temperatures and pressures that exist at large depths beneath the Earth's crust (Pennington et al., 2020).

Gneiss, slate, marble, schist, and quartzite are examples of metamorphic rocks. In building construction, slate, and quartzite tiles are employed. Marble is also valued as a building material and as a medium for sculpting. Schist bedrock, on the other hand, can be difficult to work with due to its prominent planes of weakness.

Metamorphism

Metamorphism is the process by which solid rocks undergo mineralogical and structural changes as a result of physical and chemical conditions that differ from those under which the rocks formed. Surface conditions, such as compaction, are typically eliminated from consideration. Temperature, pressure, and fluids are the most essential agents of metamorphism (Perdigão et al., 2020). Changes in the chemical environment cause two metamorphic processes: (1) mechanical dislocation, in which a rock is deformed, particularly as a result of differential stress, and (2) chemical recrystallization, in which a mineral assemblage becomes out of equilibrium due to temperature and pressure changes and a new mineral assemblage form.

Meteorite

A meteorite is any relatively small natural item from interplanetary space (i.e., a meteoroid) that survives the Earth's atmosphere and lands on the surface. In modern parlance, the word refers to comparable things that crash land on the surface of other fairly massive bodies. For example, meteorite shards have been discovered in Moon samples, and the robotic rover Opportunity has discovered at least one meteorite on Mars' surface. The largest meteorite ever discovered on Earth was discovered in Namibia in 1920 and christened the Hoba meteorite. It is 2.7 meters (9 feet) broad, weighs over 60 tons, and is composed of an iron and nickel alloy. The smallest meteorites, known as micrometeorites, ranging in size from a few 100 micrometers (m) to roughly 10 m and are formed by the population of microscopic particles that fills interplanetary space (Figure 115).

Figure 115. The 60-ton, 2.7 m-long (8.9 ft) long Hoba meteorite in Namibia is the largest known intact meteorite.

Source: https://en.wikipedia.org/wiki/Meteorite#/media/File:Namibie_Hoba_Meteorite_05.JPG.

Meteorology

Meteorology is a field of atmospheric sciences that includes atmospheric chemistry and atmospheric physics and focuses on weather forecasting. Meteorology has been studied for millennia, although real advances in meteorology did not begin until the 18th century. Weather observation networks were created across vast regions in the 19th century, resulting in moderate advances in the subject (Parkes and Morris, 2001). Previous attempts at weather prediction relied on historical data. Significant advances in weather forecasting did not occur until the latter half of the 20th century, following the elucidation of the laws of physics and, in particular, the development of the computer, which allowed for the automated solution of a large number of equations that model the weather. Marine weather forecasting is an essential branch of weather forecasting since it relates to maritime and coastal safety, and weather effects include atmospheric interactions with huge bodies of water (Figure 116).

Meteorological phenomena are observable weather events that are explained by meteorology science. Meteorological phenomena are characterized and measured by factors in the Earth's atmosphere such as temperature, air pressure, water vapor, and mass flow, as well as the changes and interactions of these variables and how they vary over time. Weather is described and predicted using many spatial scales at the local, regional, and global levels (Özyurt et al., 2021).

Figure 116. Weather scientists travel around the world all year trying to learn more about storms and how to keep people safe. Radar trucks like these are used to better understand the severity of storms and tornados.

Source: https://www.nationalgeographic.org/encyclopedia/meteorology/.

The atmospheric sciences include meteorology, climatology, atmospheric physics, and atmospheric chemistry. The interdisciplinary field of hydrometeorology is made up of meteorology and hydrology. Interactions between the Earth's atmosphere and oceans are part of a linked ocean-atmosphere system. Meteorology is used in a wide range of fields, including the military, energy production, transportation, agriculture, and construction.

Meter

The meter (Commonwealth spelling) or meter is the International System of Units base unit of length (SI), and m is the SI unit sign.

The meter is currently defined as the length of light's passage in a vacuum in 1/299 792 458 of a second. In 1793, the meter was defined as one ten-millionth of the distance from the equator to the North Pole along a great circle, implying that the Earth's circumference is approximately 40,000 km. The meter was redefined in 1,799 in terms of a prototype meter bar (the actual bar used was changed in 1889). In 1960, the meter was redefined in terms of a specific number of wavelengths of a krypton-86 emission line. The current definition was adopted in 1983 and significantly changed in 2002 to highlight that the meter is a legitimate length measure (Orion, 2019).

Microcrystalline

A microcrystalline material is a crystallized substance or rock that contains small crystals visible only through microscopic examination. There is little agreement on the range of crystal sizes that should be regarded as microcrystalline, but the extreme range of values suggested is 1 to 200 microns.

Microorganism

A microorganism, sometimes known as a microbe, is a microscopic organism that can exist as a single cell or as a colony of cells.

The existence of unseen microbial life was suspected as early as the sixth century BC in India, according to Jain literature. The scientific study of microorganisms began in the 1670s with Anton van Leeuwenhoek's observation of them under the microscope. Louis Pasteur discovered that bacteria caused food spoiling in the 1850s, disproving the hypothesis of spontaneous generation. Robert Koch discovered in the 1880s that germs caused tuberculosis, cholera, diphtheria, and anthrax (Bohle, 2015).

Microorganisms can be exceedingly diverse because they include the majority of unicellular organisms from all three domains of life. Archaea and Bacteria are the only two domains that include microorganisms. Eukaryota, the third domain, encompasses all multicellular organisms as well as many unicellular protists and protozoans that are microorganisms.

Some protists are connected to mammals, whereas others are related to green plants. There are many minuscule multicellular organisms, such as micro-animals, fungi, and algae; however these are not commonly considered microorganisms.

Microorganisms may exist in a variety of environments, including the poles, the equator, deserts, geysers, rocks, and the deep sea. Some are suited to extreme temperatures, some to high pressure, and a few, like Deinococcus radiodurans, to high radiation settings. The microbiota present in and on all multicellular organisms is likewise made up of microorganisms. There is evidence that bacteria once existed in 3.45-billion-year-old Australian rocks, the earliest direct evidence of life on Earth.

Millibar

A millibar is a unit of air pressure in the metric system, commonly used in meteorology, equal to 100 pascals, 1,000 dynes per square cm (about 0.0145 pounds per square inch), or slightly less than 1,000th of a standard atmosphere (Butler, 2016).

Mineral

A mineral or mineral species is a solid chemical substance with a very well-defined chemical composition and a specific crystal structure that occurs naturally in pure form in geology and mineralogy.

Compounds found only in living beings are typically excluded from the geological definition of mineral. However, some minerals (such as calcite) are frequently biogenic or are organic substances in the sense of chemistry (such as mellite). Furthermore, living beings frequently create inorganic minerals (such as hydroxylapatite) that are also found in rocks.

Minerals are separate from rocks, which are defined as any mass solid geologic substance that is relatively homogeneous on a big enough scale. A rock can be made up of only one type of mineral, or it can be made up of two or more different types of minerals that have been spatially separated into various phases (Britner, 2008).

Mineraloids are natural solid solids that lack a distinct crystalline structure, such as opal or obsidian. If a chemical compound exists naturally with many crystal forms, each structure is classified as a different mineral species. Thus, quartz and stishovite are two different minerals that contain the same chemical, silicon dioxide.

Figure 117. Crystals of serandite, natrolite, analcime, and aegirine from Mont Saint-Hilaire, Quebec, Canada.

Source: https://en.wikipedia.org/wiki/File:S%C3%A9randite,_natrolite,_analcime,_aegirine_300-4-2112.JPG.

The Dana classification and the Strunz classification are the two primary schemes for classifying minerals based on key chemical constituents. Silicate minerals make up approximately 90% of the Earth's crust. The native elements, sulfides, oxides, halides, carbonates, sulfates, and phosphates are other prominent mineral groups (Figure 117).

Moisture

Moisture is the presence of a liquid, most typically water, in trace levels. Water can be found in small amounts in the air (humidity), meals, and some commercial products. Moisture also refers to the amount of water vapor present in the air.

The amount of water contained in a substance, such as soil (called soil moisture), rock, pottery, crops, or wood, is referred to as its water content or moisture content. Water content is utilized in a variety of scientific and technological fields and is stated as a ratio ranging from 0 (totally dry) to the value of the material's porosity at saturation. It can be provided in either volumetric or mass (gravimetric) form (Oreskes, 2015).

Molecule

A molecule is a collection of two or more atoms that form the smallest identifiable unit into which a pure material can be divided while retaining its content and chemical properties.

The division of a substance sample into progressively smaller pieces results in no change in composition or chemical properties until parts consisting of single molecules are reached. Further subdivision of the substance results in even smaller components that usually differ in composition from the original substance and always differ in chemical characteristics. The chemical links that hold the atoms in the molecule together are broken during this stage of fragmentation.

Atoms are made up of a single positively charged nucleus surrounded by a cloud of negatively charged electrons. When atoms get close together, the electron clouds interact with one another and with the nucleus. If this contact reduces the overall energy of the system, the atoms link together to create a molecule. From a structural standpoint, a molecule is an aggregation of atoms held together by valence forces (Oreskes, 2013). Diatomic molecules are made up of two chemically linked atoms. If the two atoms are the same, as in the oxygen molecule (O_2), the molecule is called a homonuclear diatomic molecule; if the atoms are different, as in the carbon monoxide molecule

(CO), the molecule is called a heteronuclear diatomic molecule. Polyatomic molecules are molecules with more than two atoms, for as CO_2 and water (H_2O). Polymer molecules may contain many thousands of component atoms (Figure 118).

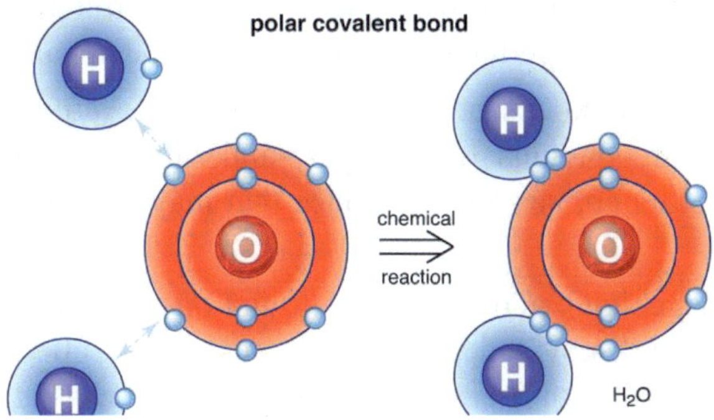

Figure 118. A water molecule is made up of two hydrogen atoms and one oxygen atom. A single oxygen atom contains six electrons in its outer shell, which can hold a total of eight electrons. When two hydrogen atoms are bound to an oxygen atom, the outer electron shell of oxygen is filled.

Source: https://www.britannica.com/science/molecule.

Molten

Molten describes an object that is reduced to liquid form by heating.

Moon

A moon is an object that orbits a planet or other non-star object. Moons can orbit dwarf planets, huge asteroids, and other entities in addition to planets. Because objects that circle other objects are also referred to as satellites, moons are sometimes referred to as natural satellites. Many artificial satellites have been launched into orbit around Earth, however these are not considered moons.

The planet or entity around which a moon revolves is referred to as its primary. Gravity keeps our solar system's planets in orbit around the Sun, and it also keeps moons in orbit around their primary stars.

Many moons formed concurrently with their main. Gravity drew dust and gas clusters together into larger and larger clumps of material. The smaller clump of material (moon) eventually began orbiting the larger cluster (primary). Other moons originated in other ways. The moon of Earth may have formed when a large object the size of Mars collided with the planet. A massive amount of material was shot into orbit around Earth as a result of the collision. This matter gradually gathered into a single big body, our moon. Other moons in our solar system were formerly asteroids, which are rock fragments too tiny to be planets. These asteroids got too close to their main and were dragged into orbit by gravity (Odenwald, 2020).

The majority of moons are composed of rock, although many also include significant amounts of ice, gas, and other substances. Europa, Jupiter's biggest moon, has an icy surface that may conceal a liquid ocean of water.

Volcanic or geologic activity can be found on some moons. Scientists, for example, have discovered volcanic plumes rising 300 kilometers (190 miles) from the surface of Io, another of Jupiter's moons (Figure 119).

Figure 119. A full moon appears when the moon is on the opposite side of the Earth from the Sun. A lunar cycle is about 29 days, so most months have one full moon. Every 2.5 years, a single calendar month has two full moons. The second full moon in a month is called a "blue moon."

Source: https://www.nationalgeographic.org/encyclopedia/moon/#:~:text=Encyclopedic%20Entry%20Vocabulary-,A%20moon%20is%20an%20object%20that%20orbits%20a%20planet%20or,are%20sometimes%20called%20natural%20satellites.

Other moons, including Earth's moon, show little or no geologic activity, despite the fact that they may have been more active in the past.

Astronomers had found 166 moons circling planets in our solar system as of 2010. Since the year 2000, 99 of them have been discovered. With 63 known moons, Jupiter has the most. There are 60 named moons on Saturn, 27 on Uranus, and 13 on Neptune. Mars only has two, and Earth only has one. Venus and Mercury do not have moons.

Six more moons in our solar system orbit dwarf planets. Dwarf planets are objects that resemble planets but do not meet the full definition of a planet. Pluto is the most well-known dwarf planet. Pluto is home to three moons (Neeck and Volz, 2011). Many more moons circle smaller things in our solar system. Because moons are so small, none have been identified outside the solar system, although there are possibly billions of moons in the cosmos.

Moraine

A moraine is the material left by a moving glacier. Typically, this material consists of soil and rock. Similarly, to how rivers move trash and silt that eventually accumulates up to produce deltas, glaciers transport dirt and stones that eventually stack up to form moraines (Figure 120).

Figure 120. Rubble covers the glacier that runs down the slope of El Tronador, Chile. The debris that accumulates at the bottom, or snout, of a glacier is called the end moraine.

Source: https://www.nationalgeographic.org/encyclopedia/moraine/#:~:text=Encyclopedic%20Entry%20Vocabulary-,A%20moraine%20is%20material%20left%20behind%20by%20a%20moving%20glacier,build%20up%20to%20form%20moraines.

Imagine yourself with a toy bulldozer on a lawn with a pile of dried leaves all over it to get a better concept of what moraines are. When you run the bulldozer over the leaves, some are pushed aside, some are pushed forward, and some form intriguing patterns on the grass. Consider the patterns and piles of pushed-away leaves—moraines—that extend for kilometers across the Earth (Movva et al., 2012).

Moraines can only be found where there are or were glaciers. Glaciers are massive, flowing rivers of ice. Glaciers shape the landscape in a process called glaciation. Glaciation can affect the land, rocks, and water in an area for thousands of years. That is why moraines are often very old.

Motion

Motion is the phenomena in which an object's position varies over time. Displacement, distance, velocity, acceleration, speed, and time are all mathematical words used to describe motion. A body's motion is observed by connecting a frame of reference to an observer and measuring the change in location of the body relative to that frame as time passes. Kinematics is the discipline of physics that studies forces and their effects on motion; dynamics is the branch that studies forces and their effects on motion.

An object is said to be at rest, motionless, immobile, stationary, or to have a constant or time-invariant location with respect to its surroundings if it does not change relative to a given frame of reference. Absolute motion cannot be determined since there is no absolute frame of reference (Müller et al., 2019). As a result, everything in the cosmos can be thought to be in motion.

Plate motion can be relative or absolute. Relative plate motion describes the motion of one tectonic plate relative to another. Absolute plate motion describes the motion of one plate relative to a fixed reference system. Plate motion can be described by a pole of rotation and an angular velocity about this pole.

Muddy

Mud is a mixture of soil, loam, silt, or clay and water. It commonly appears after a rainstorm or near a body of water. Over geological time, ancient mud deposits solidify to form sedimentary rock such as shale or mudstone (generally called lutites). When mud deposits form in estuaries, the resulting layers are known as bay muds.

A mud volcano, also known as a mud dome, is a type of landform formed by the eruption of mud or slurries, water, and gases. Mud volcanoes can occur as a result of a variety of geological processes. Mud volcanoes are not true igneous volcanoes because they do not emit lava and are not always fueled by magmatic activity (Metz, 2008). Mud volcanoes can be as little as 1 or 2 meters high and 1- or 2-meters broad, or as large as 700 meters high and 10 kilometers wide. Mud-pots are little mud exudations that exude mud. Mud volcanoes produce mud primarily when hot water heated far beneath the Earth's surface begins to mix and blend with subsurface mineral deposits, resulting in the mud slurry exudate (Figure 121).

Figure 121. A group of mud volcanoes in Gobustan, Azerbaijan.

Source: https://en.wikipedia.org/wiki/Mud_volcano#/media/File:Gobustan_State_Reserve_04.png.

N

Natural Gas

Natural gas is formed beneath the Earth's surface and utilized as a fossil energy source. There are various components found in natural gas methane being the largest component. Methane has four hydrogen atoms and one carbon atom. Other than methane, natural gas contains small amounts of natural gas liquids, also called hydrocarbon gas liquids and non-hydrocarbon gases. Natural gas is used as a fuel (Duffy, 2014). Research points out that natural gas was formed when the remains of plants and animals built up in thick layers in ocean floors and Earth's surface. The formed layers were then buried under rock, silt, and sand. These remains are rich in hydrogen and carbon. Heat and pressure act on these remains resulting in the formation of natural gas, oil, and coal (Figure 122).

Figure 122. Natural gas is a natural source of energy.

Source: https://www.nationalgeographic.org/encyclopedia/natural-gas/.

Nebula

Nebula is a term used to refer to a distinct body of inter stellar cloud. Molecular clouds, helium, hydrogen, and cosmic dust from the nebula. Initially, the term nebula was used to refer to diffused astronomical objects such as galaxies located beyond the Milky Way. For instance, the Andromeda galaxy was once called the Andromeda Nebula before the confirmation of true galaxies. It was discovered that most nebulae are associated with starts and that start lights illuminate the nebulae. The type of light spectra produced by nebulae was used in categorizing nebulae (Clowes, 2010). Some nebulae are hundreds of light-years in diameter and have vast sizes. If a nebula is visible to the human eye from Earth it would appear very large, but when compared to close observation, it would be less bright. The brightest nebula is the Orion nebula, and it occupies an area twice the angular diameter of the full moon. It is visible to the human eye when viewed from Earth. Nebulae are less dense than any vacuum created on Earth but are denser than the space surrounding them. It is said that if a nebular cloud was of the same size as the Earth, it would have a total mass of a few kilograms. Embedded hot stars cause fluorescence making many nebulae visible. Some nebulae are not visible but can be detected by using special filters (Figure 123).

Figure 123. The nebula takes the form of a cloud or fog.

Source: https://www.nasa.gov/mission_pages/chandra/images/helix-nebula.html.

Neutron

A neutron is a subatomic particle that does not carry any charge. Neutrons are found in the nucleus of every atom in exception of hydrogen. The rest mass of a neutron is equal to 1.67493×10^{-27} kg. The mass of a neutron is greater than that of a proton. Neutrons and protons account for the mass of an atom. It was discovered that neutrons are not true elementary particles. Neutrons contain elementary particles called quarks. The strong force helps keep both neutrons and protons in the nucleus (Clary and Wandersee, 2011). The strong force influences the behavior of quarks that make up a neutron. Neutrons are used in nuclear reactors to bombard other elements resulting in the production of energy. Neutrons are also produced during nuclear reactions. The production of neutrons helps maintain a chain reaction. This led to the development of atomic bombs. Certain materials are able to absorb neutrons in their nuclei. These materials are used to control nuclear reactions in nuclear reactors. Neutrons have been studied and have been very useful in giving more information of the nucleus.

Nickel

Nickle is a naturally occurring chemical element with a silvery-white appearance. When compared to iron, nickel is harder and tough. Nickel is a ferromagnetic material and found in group 10 of the periodic table. Nickel is said to be very resistant to corrosion and oxidation. Nickel has extensive applications in industrial and domestic areas as an alloy or a pure metal. Some types of nickel are found in meteoric and terrestrial deposits. These are the elemental nickel which tends to occur with iron.

Nitrate

Nitrate is an oxoanion of nitrogen formed when nitric acid losses a proton. Nitrates have a pH of 7.3. Nitrates are principal species and are considered to be reactive members of nitrogen species. They are also monovalent inorganic anion. Nitrates are conjugate bases of nitric acid. Inorganic aqueous nitrate solutions or n.o.s readily ignite when they come into contact with organic materials (Cossu et al., 2010). Nitrates are naturally non-combustible though when put with combustible materials will accelerate their combustion. Nitrates are relatively harmful to human beings. They find their way to the body through inhalation, contact or ingestion. When burnt, nitrates produce toxic, corrosive, or irritating gases. Water used in diluting nitrates can cause pollution when released to the environment.

Nitrogen

Nitrogen is a non-metallic element gaseous in nature. It is group 15 of the periodic table. Large amounts of nitrogen are in the Earth's atmosphere.

Nitrogen is colorless, odorless, and tasteless. When ranked alongside other elements, nitrogen is ranked sixth in cosmic abundance. There is about 78% of nitrogen in the atmosphere hence the primary source of nitrogen in industries (Cerase, 2019). Other than nitrogen, the atmosphere contains small amounts of nitric acid, nitrogen oxides, ammonium salts and ammonia. Nitrogen can also be found in gases released from volcanoes, mineral springs, mines, nebulae, some stars, the Sun, and meteorites (Figure 124).

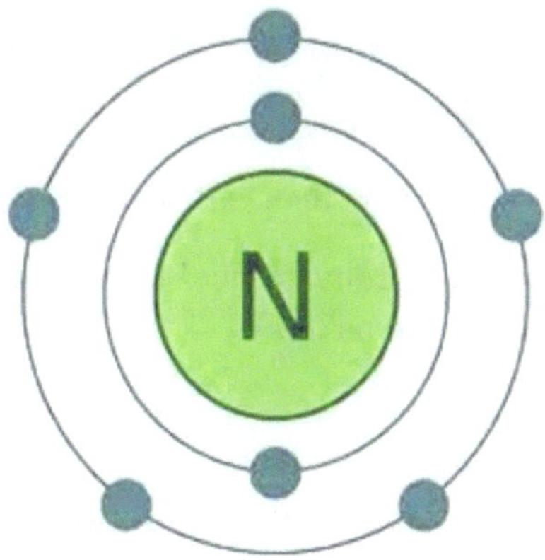

Figure 124. Nitrogen is among gases in the atmosphere.

Source: https://chemindigest.com/nitrogen-life-essential-element/.

Northern Lights

Northern lights are also called aurora Polaris or polar lights. They are a natural display in the sky. They are mostly observed around the Antarctic and Arctic regions (high altitude areas). Northern lights display dynamic patterns of brilliant lights appearing as dynamic flickers, spirals, rays, or curtains covering the sky. Solar winds disturb the magnetosphere resulting in northern lights. The disturbances occur in the magnetospheric plasma and they cause an alteration of the trajectories of charged particles. Some of the charged particles include protons and electrons (Britner, 2008). They precipitate into the exosphere or thermosphere. When precipitated to the atmosphere, these particles cause excitation and ionization of atmospheric constituents which emit light with varying complexity and color. Northern lights occur within bands. These bands influence the form of the northern lights. The amount of acceleration imparted to precipitating particles also influence the form of the northern lights. The auroral zone is the common band in which northern lights occur (Figure 125).

Figure 125. The northern lights are also called aurora Polaris.

Source: https://www.nbcnews.com/science/science-news/scientists-searched-elusive-sounds-northern-lights-rcna2840.

Nuclear

Nuclear is a term mostly used when relating to particles found in the nucleus. These nuclear particles are naturally occurring in nature, and over the years these particles have been used in the generation of electricity. This form of power is called nuclear power. Nuclear power is generated through nuclear fission. The process results in the production of heat and electricity which is harnessed and used to power several America homes (Butler, 2016). Nuclear power is preferred compared to other forms of energy as it requires low-carbon energy and is very reliable. The use of nuclear energy has been embraced as a method of dealing with pollution due to carbon emission and global warming.

Nuclear Fission

Nuclear fission is a form of nuclear reaction characterized by the nucleus of an atom-splitting into two or more smaller nuclei. This nuclear reaction results in the production of large amounts of energy and gamma photons. Energy produced from the reaction can meet the energetic standards of radioactive decay. Initially, nuclear fission was done using light elements, and eventually heavy elements are used in these reactions. Nuclear fission was named using an analogy of biological fission of living cells (Bohle, 2015). When heavy elements are used, the reaction is an exothermic reaction causing the production of large amounts of energy as either kinetic energy or electromagnetic radiation. Nuclear reactors produce energy through nuclear fission. Nuclear fission is done by having a neutron being fired towards the nucleus of an atom. The fired neutron possesses and energy that causes the nucleus of the atom to split. The resultant nuclei are light compared to the initial atom (Figure 126).

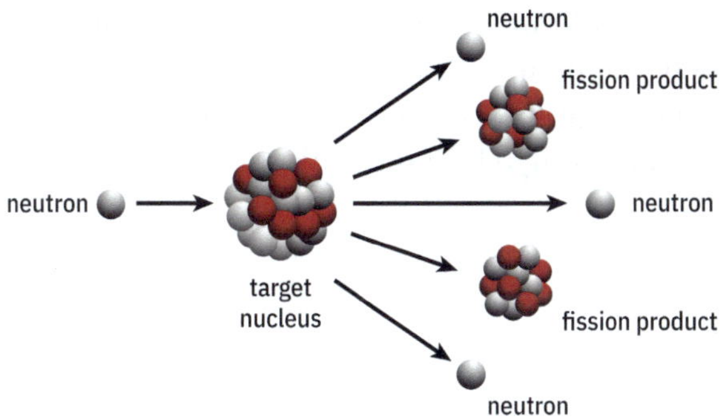

Figure 126. Nuclear fission involves the splitting of atoms.

Source: https://www.atomicarchive.com/science/fission/index.html.

Nuclear Fuel

Nuclear fuel is a fuel used in nuclear reactors. The nuclear fuel is enriched to attain a critical mass allowing it to sustain a nuclear chain reaction. Radioactive elements such as plutonium – 239 and uranium – 235 are commonly used nuclear fuels. Nuclear fuels are fissile. The nuclear fuel cycle outlines the processes involved in obtaining, refining, and using nuclear fuels. As a nuclear fuel, uranium – 235 can be used at different concentrations (Zhuang et al., 2019). There are fuel reactors that use natural uranium while other use enriched fuels. Fast breeder reactors use plutonium as their fuel. They produce their plutonium and use in reactors. This plutonium fuel contains significant amounts of uranium – 238. It is at times recycled and used in thermal reactors as fuel.

Nuclear Fusion

Nuclear fusion is the opposite of nuclear fission in that elements combine to form a new element. Nuclear fusion results in the production of energy. Nuclear fusion mostly involves elements with low atomic numbers. A good illustration of nuclear fusion is the power fueling the Sun and stars. The Sun is a ball of hydrogen gas. Hydrogen molecules combine under pressure to form a neutron and helium. The amount of energy produced during nuclear fusion is tremendous. The reactions are therefore called thermonuclear reactions (Yu and Gong, 2012). Nuclear reactors do not use nuclear fusion in the generation of energy because there is no known material that can be used to confine nuclear fusion reactions. The high amounts of temperature could cause a material to move to the fourth state of matter called plasma. There is so much potential with nuclear fusion as there is no limit in the amount of fusion that can occur. This is different for nuclear fission and the fission process can be stopped (Figure 127).

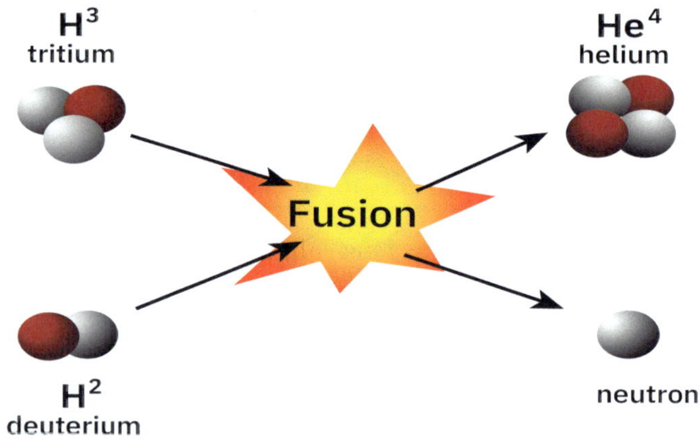

Figure 127. Nuclear fusion involves the combining of elements.

Source: https://www.atomicarchive.com/science/fusion/index.html.

Nucleus

The nucleus is the central part of atoms. It is composed of neutrons and protons. Protons are positively charged, while neutrons do not carry any charge. For this reason, the nucleus of an atom is said to be electrically positive. The mass of protons and neutrons in the nucleus contribute a large mass of an atom. The term mass number is used to refer to the total number of neutrons and protons in the nucleus (Witze, 2013). Nuclear reactions take effect the nucleus allowing the formation of new elements. The nucleus is very small in size and is surrounded by electrons. Electrons move around the nucleus in orbits. The probable location of the nucleus is called orbitals. The mass of electron cannot compare to that of protons and neutrons. The mass of the nucleus is small compared to the individual masses of protons and neutrons. This is because some mass is lost as energy when protons and neutrons combine to form the nucleus (Figure 128).

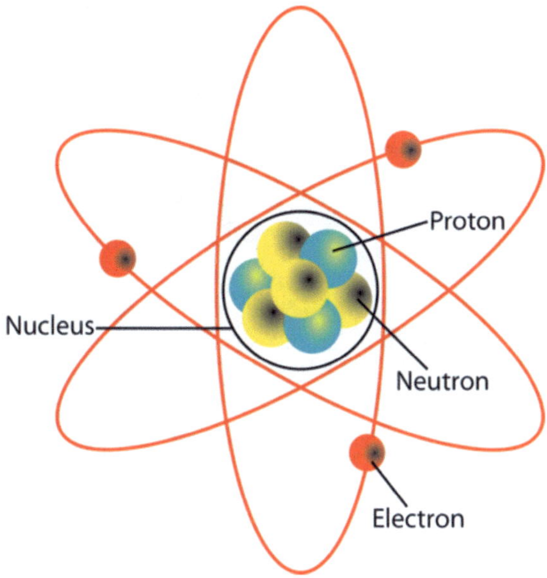

Figure 128. The nucleus contains protons and neutrons.

Source: https://socratic.org/questions/what-two-particles-are-found-in-the-nucleus-of-an-atom.

Nutrient

Nutrients are chemical substances naturally found in soils and living things. Nutrients are a necessity for animals, plants, people, and other organisms. Organisms use nutrients in breaking down food to produce energy and are used in all other processes in the body of an organism. These processes include breathing, repair, and growth. People get most of their nutrients from plants. Plants get nutrients from water and the soil. Some living things are able to manufacture their own food. They are called autotrophs. For them to make food, they need nutrients such as oxygen, hydrogen, and carbon. Some of the nutrients needed by plants include sulfur, magnesium, calcium, potassium, phosphorous, and nitrogen (Wilkinson et al., 2020). By obtaining these nutrients, plants, and autotrophs are able to create their own nutrients. Human being synthesizes some nutrients. Autotrophs play a vital role in the supply of nutrients as most living things rely on them. Certain nutrients are considered essential. The essential nutrients are those that cannot be synthesized by the human body. They can only be gotten from water and plants. Some rocks are very rich in nutrients. When the weather, they release some of the nutrients to soil and water (Figure 129).

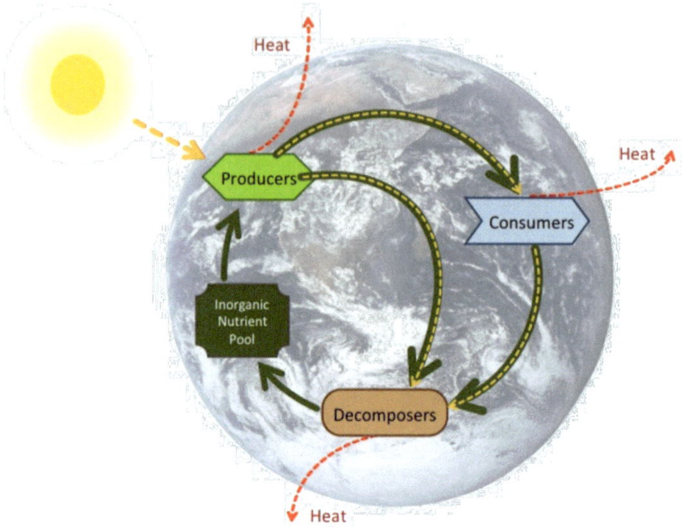

Figure 129. Nutrients are naturally found in soil.

Source: https://www.khanacademy.org/science/biology/ecology/biogeochemical-cycles/a/introduction-to-biogeochemical-cycles.

O

Ocean Floor

The ocean floor also called the seabed. The shape of the ocean floor is called the bathymetry. The shape of the ocean is determined by a process called plate tectonics. Ocean beds are mostly covered by sand. Oceans cover large areas of land. Ocean beds have Earth cover. The outer rocky layer of the Earth forms part of the plate tectonics.

Figure 130. Ocean floors may carry living things such as coral reefs.

Source: https://www.scientificamerican.com/article/new-xprize-challenge-map-the-ocean-floor/.

Ocean beds may change or increase in sizes or changes as a result of the convectional currents in the molten mantle (Wallace-Wells, 2019). These movements cause the tectonic plates to move about the Earth a few centimeters. These interactions occur at the edges of the plates result in the features in the ocean floors. As the plates shift it may slide past each other, diverge, or converge. When the plates converge, deep ocean trenches are formed. When they diverge, it results in the creation of new floor crust, hydrothermal vents, underwater volcanoes, and mid-ocean ridges (MORBs). The faults that connect two areas where plates are diverging or converging are called the transform boundaries. In most cases, the edges of the continental boundaries form zig-zag patterns. Ocean grounds also create breeding ground for sea animals (Figure 130).

Oceanic

Oceanic is a term used when relating to the ocean. The term is also used when relating to certain aspects of oceans. For instance, oceanic climate. This refers to the climate of the ocean. Oceanic currents refer to currents that cause water to move. Ocean currents influence the land breeze and also vegetation in the ocean. Oceanic storms refer to storms exhibited in the sea.

Orbit

Orbit is the term used to refer to a regular path that objects in space take around another. For instance, planets revolve around the Sun round a circular path called orbits. Some of the objects that orbit around the planet include the moon and satellites. Several planets are known to have moons which orbit around them for a given period of time. Satellites are examples of man-made objects that orbit the Earth. Other than planets, asteroids, comets, and other objects orbit the Sun. These objects orbit round the Sun close to or along an imaginary flat surface. The term ecliptic plane is used to refer to the imaginary surface. Orbits have different shapes. All orbits are similar to an oval, meaning they are an ellipse (Toth et al., 2014). For this reason, orbits are said to be elliptical. In the case of planets, they have circular orbits. In the case of comets, their orbits are said to be squashed or highly eccentric. They mostly look like ellipses rather than circles. Orbits also have different diameters. For instance, the moon orbits the Earth at a different distance compared to satellites. The orbit distance for the moon round the Earth also changes. In some cases, the moon orbits close to the Earth, and in other cases it orbits far from the Earth. In the case of the Earth, the closest orbiting point of a satellite is called a perigee while the furthest point is called the apogee (Figure 131).

For planets, the point where the planet orbits closest to the Sun is called the perihelion, while the furthest point is called the aphelion. During summer, the Earth reaches its aphelion. The term period is used to refer to the time taken for a satellite to make a full orbit. Earth makes a full orbit in the period of a year. The orbital plane makes an angle of inclination compared with the Earth's equator.

Figure 131. Orbits are circular paths in which planets revolve around the Sun.

Source: https://www.livescience.com/planets-orbit-same-plane.

Ore

Ore is a term used to refer to a natural aggregation of minerals. These minerals can be mined, be processed, and sold. Initially the term ore was limited to refer to metallic mineral deposits. Currently, the term is also used to refer to non-metallic minerals. The number of identified mineral species are about 2,800. Out of the 2,800 minerals 100 are considered ore minerals. They include siderite, limonite, magnetite, and hematite. These ore minerals are the principal sources of lead, zinc, galena, sphalerite, copper, chalcocite, bornite, and iron (Tollefson, 2015). Disseminated deposits have been found to carry gold, molybdenum, and copper. These deposits can be found in large volume of rock. Massive sulfide deposits carry zinc, lead, and copper. Some of the deposits are formed due to volcanic exhalations that produce precipitation. They are also formed by metasomatic replacement. There are no records of deposits made of single ore mineral. In most cases, the ore is mixed with valueless minerals and rocks. They are collectively known as gangue. The gangue and ore are mined together. They are harvested from a host rock mass by manual or mechanical means. Various operations are used in separating the ore from the gangue. These processes are collectively called ore dressing or mineral processing (Figure 132).

Figure 132. An ore is a naturally occurring material.

Source: https://www.britannica.com/science/ore-mining.

Organic

Organic is a term used to refer to a class of chemical compounds or substances comprised of those derived or existing in animals and plants. Carbon compounds have also been categorized as organic substances. The term organic is also used to refer to naturally occurring substances. For instance, if a plant has been grown in soil that does not contain any synthesized substances, it is referred to as organic plants.

Organism

The term organism is used to refer to an organic living system able to function as an individual entity. The basic unit of an organism is a cell. There are different organisms which can be classified as fungi, plants, or animals. Every organism has the ability to respond to stimuli, grow, reproduce, and develop. Some organisms have specialized organs and tissues. There are unicellular organisms and multicellular organisms (Thessen, 2016). Eukaryote and prokaryote are the two main groups of unicellular organisms. There are some living things that portray characteristics of organisms (they can respond to stimuli) but are not able to reproduce or grow. These living things are not considered to be organisms. They include viruses (Figure 133).

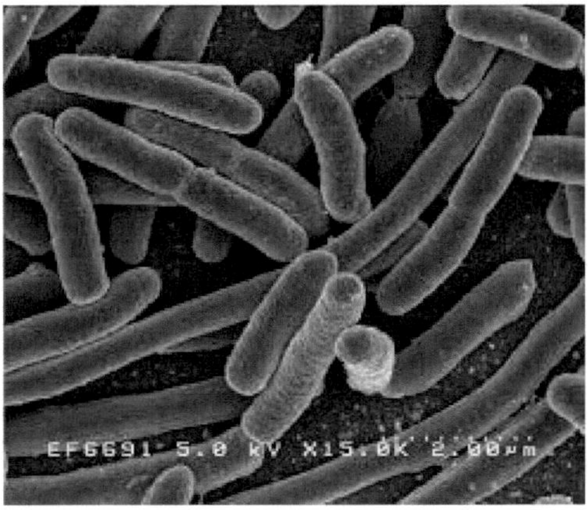

Figure 133. An organism is a living structure.

Source: https://en.wikipedia.org/wiki/Organism.

Orifice

Orifice is an opening in features such as mountains or hills. These openings could be as a result of volcanic activity or landslides. Orifice allows outflow of water or magma from beneath the Earth's surface. Pressure within the Earth could cause displacement of soil creating an opening allowing the outflow of hot air.

Oxide

The term oxide is used to refer to a chemical compound containing an element and at least one oxygen atom in its chemical formula. An oxide is a dianion of oxygen or a molecular ion of oxygen. There are different kinds of oxides they include metal oxides. Metal oxides have an anion of oxygen in the oxidation state of –2. There are solid oxides in the Earth's crust. For this results in elements being oxides by oxygen in water or air. There are cases where pure materials have an oxide coating. This is well demonstrated by aluminum foil that develops a thin skin of aluminum oxide. This skin helps to protect that aluminum foil from corrosion. There are cases where elements for multiple oxides. They are different depending on the amounts of elements that combine with oxygen. Aluminum, lithium, titanium, silicon, nitrogen, iron, and carbon are examples of such elements (Telling et al., 2017. For these elements, the oxides are distinguished by specifying the elements oxidation number or specifying the number of atoms involved. There are a variety of oxide structures, there are molecules with crystalline and polymeric structures. Oxides may be solids or liquids at standard conditions (Figure 134).

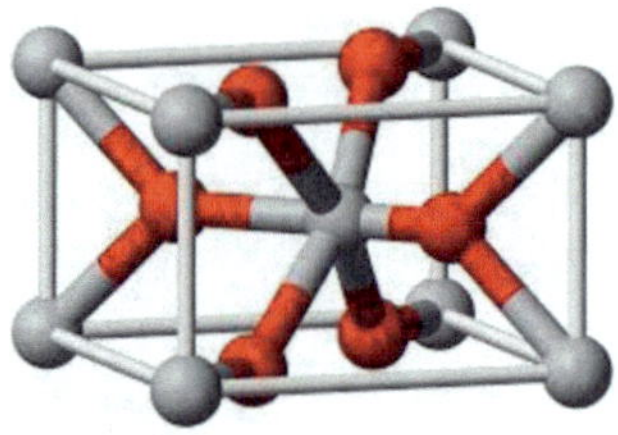

Figure 134. An oxide is a compound formed when oxygen combines with an element.

Source: https://en.wikipedia.org/wiki/Oxide.

Oxygen

Oxygen is a naturally occurring gas in the atmosphere. It is also a chemical element represented by the symbol O. It is of atomic number 8. In the periodic table, oxygen is under the chalcogen. Oxygen is a highly reactive non-metal. It readily forms oxides with several elements and is therefore an oxidizing agent. Oxygen can form oxides with some compounds. Among the naturally occurring elements on Earth, oxygen is the most abundant. In the universe, it is the third most abundant element.

Two atoms of oxygen at standard temperature and pressure bind to form dioxygen. Oxygen is a colorless and odorless diatomic gas. As a diatomic gas, oxygen constitutes about 20.95% of the Earth's atmosphere (Showstack, 2008). This percentage has changed over the years. About half of the Earth's crust is made up of oxygen in the form of oxides. Photosynthesis has actively been involved in the continuous replenishment of oxygen in the atmosphere.

During photosynthesis, plants use energy from the Sun and carbon dioxide (CO_2) and water in the environment is used to produce oxygen. Oxygen is continuously replenished in the atmosphere as oxygen is a chemically reactive element and its amount would be exhausted if it was not continuously replenished. The ozone is an allotrope of oxygen. The ozone protects the Earth's atmosphere from ultraviolet (UV) radiation.

Ozone

Ozone also called trioxygen, is an inorganic molecule having a chemical formula of O. The ozone is a pale blue gas known for its pungent smell.

Compared to other allotropes of oxygen such as diatomic allotrope O, ozone is the most stable. Dioxygen is involved in the production of ozone by electrical discharges within the Earth's atmosphere and action of UV light. There are very low concentrations of ozone in the earth's atmosphere while high concentrations in the ozone layer of the stratosphere. Ozone in the ozone layer is useful in absorbing most of the Sun's UV radiation. Chlorine contributes to ozone odor. This makes it easily detectible even when small amounts are in the air. In 1865, ozone's structure was determined. Ozone is weakly paramagnetic and has a bent structure (Schmitz, 2011). Ozone is a pale blue gas in standard condition. In cryogenic temperatures, ozone condenses to dark blue liquid. It then becomes a solid, violet-black in color. At elevated temperatures, liquid ozone decomposes explosively. This is why ozone is used at low concentrations (Figure 135).

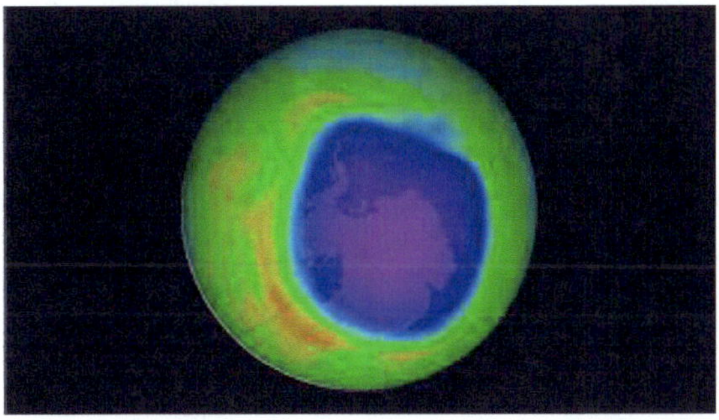

Figure 135. Ozone is an allotrope of oxygen.

Source: https://www.livescience.com/ozone.html.

Ozone Hole

Ozone hole is the term used to refer to the depletion region in the ozone layer. The depletion is as a result of chemical reactions involving bromine and chlorine that causes the ozone layer to be rapidly and severely destroyed. A map of total column ozone is used to determine the area of the ozone hole. Ozone hole is calculated from the area of the Earth enclosed by a line having 220 Dobson units (Schmitz, 2012). It was originally established that ozone has values greater or equal to 220 Dobson units. During historic observations, there were no records of ozone with values less than 220 Dobson units. When values less than 220 Dobson units were recorded, it is interpreted as an ozone hole as ozone was lost due to reactions between bromine and chlorine compounds (Figure 136).

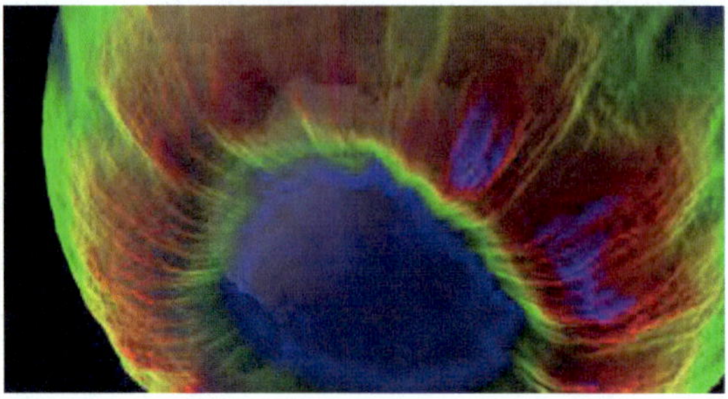

Figure 136. Ozone hole is a deleted region in the ozone layer.

Source: https://www.pbs.org/show/ozone-hole-how-we-saved-planet/.

Ozone Layer

The ozone layer also called the ozone shield is a region of the Earth's stratosphere useful in absorbing the Sun's UV radiation.

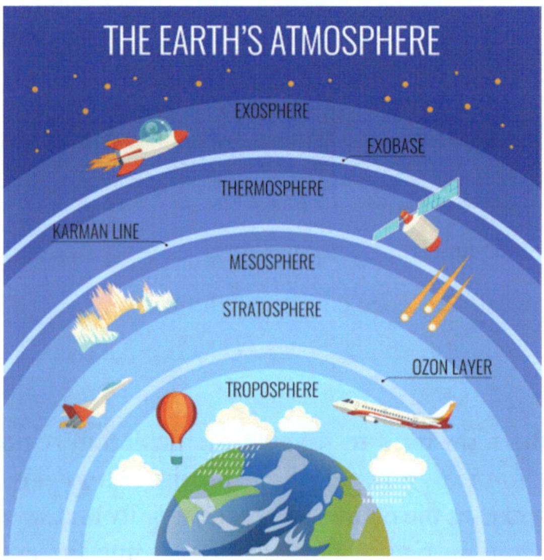

Figure 137. The ozone layer protects the Earth from harmful UV rays.

Source: https://www.business-standard.com/about/what-is-ozone-layer.

The ozone layer is made up of high concentrations of zone compared to other parts of the atmosphere. Although the ozone layer contains high concentrations of ozone, the concentration of ozone is small compared to other gases. Amounts of ozone in the ozone layer is less than 10 parts per million of ozone which is larger than the average concentration of ozone on Earth's atmosphere. The ozone layer is found in the lower part of the stratosphere. The ozone layer is about 15 to 35 km from the Earth (Rao, 2012). The thickness of the ozone layer varies geographically and seasonally. The continuous release of air pollutants from industries has led to the depletion of the ozone layer. This has caused the increased amount of UV radiation on the Earth's surface (Figure 137).

P

Perfect Gas

Perfect gas is a gas that adjusts, in actual form, to a specific, idealized connection between pressure, capacity, and temperature known as the standard gas law. Such a connection for a substance is known as the equation state and is adequate to portray its gross conduct. Different gases, for example, hydrogen, nitrogen, oxygen, noble gases, denser gases like carbon dioxide (CO_2) and combinations like air, can be considered as ideal gases across reasonable tolerances over an impressive perimeter range on standard temperature and tension (Rivet and Kastens, 2012). Notwithstanding, the gas particles must possess zero capacity and they need to display no attractive powers at all toward one another. Since none of those factors can be valid, a perfect gas cannot exist (Figure 138).

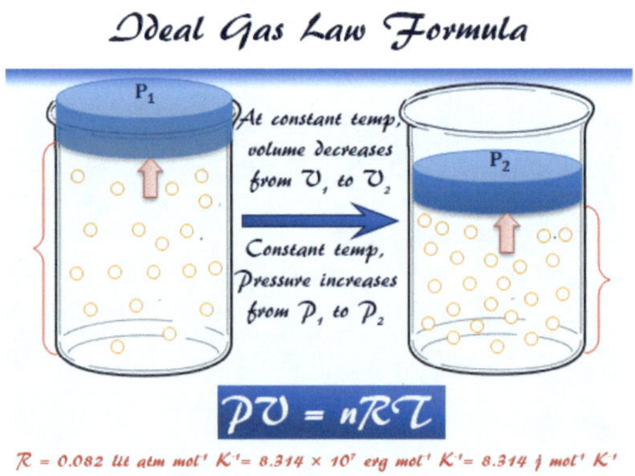

Figure 138. Perfect gas occupies less space than other gases.

Source: https://www.priyamstudycenter.com/2019/04/ideal-gas-law.html.

Period

Period, in geography, is the primary unit of the geologic timeframe; during these ranges of time unique rock systems were shaped. Initially, the serial idea of describing periods was a comparative one, beginning from the context of comparing stratigraphic arrangements and the proof gotten from paleontological inquiries.

With the invention of radiometric dating techniques, absolute ages for different periods can be identified. Similar to the periodic chart, the Geologic time frame is one of those astounding human undertakings to gain proof from astronomy, history, rocks, fossils, and political issues into an outline which makes the entire process enjoyable.

Periods are units used to depict parts of the Earth's experiences, however where these start and end depend on a subjective interpretation of the rock history. The titles of these units record somewhat the historical backdrop of their portrayal as well (Pyle et al., 2018). Isolating geologic time has likewise been the reason for various professional run-ins between geologists, such as the Great Devonian discussion and accordingly the Devonian period is christened after the Devon face-saving rocks.

Permeable

Permeability is characterized as the capacity of a liquid of unit consistency moving through a unit cross-part of the medium in unit time across the activity of a unit pressure angle. Permeability is a proportion of the ease of movement of fluids or gases or particular chemicals through the material. Permeability is defined by providing ahead and deciding the penetration of entrance or the measure of fluid or gas passing through the sample. Estimations on heads of pressure can be enhanced by capillarity and porosity research.

Permeability estimates for porous rocks are regularly estimated by geologists through the pore pressure oscillation strategy. This is no doubt the best strategy to use for different materials also, yet some adjustment might be required based upon the material used. A computational technique for determining the penetrability of texture preforms dependent on the Lattice Boltzmann strategy has been framed and demonstrated to be solid at predicting successful preform permeability dependent on the quantity of texture layers within the preform, shear point, as well as the compaction power (Press, 2008). This or a comparative program might be helpful for determining permeability in a particular situation. Permeability is significant for two reasons: the parameter defines how quickly deleterious products might enter the material and how effectively material can be filtered from it.

Petroleum

Petrol is a naturally existing liquid found beneath the Earth surface which can be changed into fuel. Petrol is a non-renewable energy source, implying that it has been made by the disintegration of natural matter millions of years in the making. Petrol forms when enormous amounts of dead life forms basically zooplankton and green growth under sedimentary stone are exposed to immense heat and tension (Plotnick et al., 2014). Petrol is utilized as fuel to control vehicles and machinery, just as being changed over into plastics and different materials. Since most of the world depends on petrol for some products, the petrol business is very influential and bears a significant effect on global legislative issues and the economy (Figure 139).

Figure 139. Petroleum is used to fuel vehicles and other machines.

Source: https://www.britannica.com/quiz/petroleum-fact-or-fiction.

The abstraction and handling of petrol, and along these lines, its accessibility, is a significant driver of the global economy and international affairs. Probably the biggest organizations on the planet are concerned with the extraction and handling of petrol, and numerous different organizations make products that are oil based, such as plastics, composts, vehicles, and aircraft, for instance. Asphalt, which is utilized to pave expressways, is produced using petroleum. Automobiles that drive on expressways are made of materials gotten from petrol and powered by fuels gotten from petrol.

Phanerozoic

Phanerozoic is the contemporary topographical age. It started roughly 541.0 million years prior. It is forerun by the Precambrian eons. The Phanerozoic started with the Cambrian Eruption, which was the huge biodiversification of multicellular creatures that formed during the late Precambrian (Pennington et al., 2020). The Phanerozoic is subdivided into three periods: the Cenozoic or "Period of Recent Life"(current time), the Mesozoic or "Middle Life," and the Paleozoic or "Ancient Life." The identification of the Phanerozoic times mirrors the series of life emerging over the ages, beginning with the original bacterial lifeforms and marine creatures, followed by dinosaurs and early vertebrates, through to people and the other current creatures.

There are sure significant events that have happened intermittently during the Phanerozoic (Metz, 2008):

- Mass extinctions are eliminations where huge numbers of species vanish. Throughout the Phanerozoic, there have been five significant extinction periods and a few more modest terminations. The purposes behind the extinctions differ yet may generally cause shifts in Earth's natural conditions, making it hard for certain species to thrive.
- Significant glaciations-The Earth has intermittently gone through immense glaciations, which can cause worldwide cooling and lowers the ocean level.
- Magnetic pole inversions The Earth's "switch" sporadically, with the North Pole turning into the South Pole and "switch" back onto the successive reversal (Perdigão et al., 2020).
- Climatic changes. The Earth's environment has moved from hot and cool, dry, and wet since the Earth's commencement. This has been a characteristic environmental change process, not to be mistaken for anthropogenic, or human-based, environmental change.

Phase

A recently formed terrestrial planet, like the Earth or Venus, passes through four unique phases of development: Differentiation, Cratering, Deluge, and Surface Evolution. Researchers believe that rocky planets formed by clustering together from residue and gas into hot masses of liquid metal and rock a few billion years ago. In space science, phase is any of the changing appearances of heavenly bodies as various measures of its loop are seen (from Earth, traditionally) to be enlightened by the Sun. Similarly, the Moon passes through four fundamental stages: new, first quarter, full, and final quarter.

Phosphate

A type of phosphoric acid containing phosphorus. Inside the body, phosphates are present in the teeth and bones. Phosphates might be utilized to treat high degrees of calcium inside the blood. Adding or eliminating phosphate chemical groups might influence the manner in which proteins function in the body (Figure 140).

Figure 140. Phosphate is a naturally occurring mineral.

Source: https://www.gardeningknowhow.com/garden-how-to/soil-fertilizers/rock-phosphate-fertilizer.htm.

In science, a phosphate is an anion, salt, practical group, or ester got from a phosphoric acid. It most ordinarily implies orthophosphate, a subsidiary of orthophosphoric acidic HPO (Parkes and Morris, 2001). Orthophosphates are particularly common among the different phosphates in light of their critical roles in biology and biogeochemistry, including their economic significance for agribusiness and industry. The expansion and expulsion of phosphate groups (phosphorylation and dephosphorylation) are key stages in cell digestion.

Photosphere

The photosphere is the bottom level of the solar environment. It is basically the solar "surface" that is seen when checking out the Sun in "white" (for example regular, or visible) light. When we check sunspots and faculae (brilliant small cloud-like elements) we are noticing them in the photosphere. The photosphere is the discernible "surface" of the Sun. There is some variety in temperature from one spot to another within the photosphere – for instance, sunspots are generally cool areas with heat levels as low as 3,000–4,500°C (5,400 to 8,100°F).

The Sun's photosphere resembles the Earth's crust somehow or another. The photosphere and crust are several miles thick. The highest point of the covering is the outer layer of the Earth. If we could be on the moon and examine the Earth, we would check its surface or crust (Özyurt et al., 2021). Assuming we review the Sun, we would see the photosphere. We can stroll around on the outer layer of the Earth but not really for the Sun. The photosphere is not solid. You would fall directly through, what is more, the Sun is extremely hot (around 5,000°C). Most stars contain photospheres. Thus, the solar photosphere is something you would find in practically any star in the universe.

Physics

Physics is a scientific study dealing with the form of matter and connection between the basic components and the associations between the basic constituents of the visible universe. In the widest sense, physics (from the Greek physikos) examines all parts of nature on the macro and submicroscopic levels. Its extent of study includes not just the conduct of objects within the action of particular forces, yet additionally the nature and beginning of gravitational, electromagnetic, and atomic power fields. Its definitive objective is the formation of a couple of comprehensive rules that unite and clarify all such different physical peculiarities.

Until recently, physics, and natural philosophy were reciprocally used for the science whose point is the revelation and definition of the primary laws of nature. When modern sciences evolved and became progressively specific, physics came to mean that aspect of physical science excluded from cosmology, geography, and psychophysics (Orion, 2019). Physics assumes a significant part in every one of the basic sciences, be that as it may, and all such fields have branches in which physical laws and estimations have exceptional accentuations, bearing such names as astronomy, geophysics, biophysics, and even psychophysics. Physics can be characterized as the study of matter, movement, and energy. Its laws are usually communicated with economy and accuracy in the form of arithmetic.

Physicist

A physicist is a researcher who focuses on the field of material science, which incorporates the interactions between matter and energy, including at all length and time spans in the physical world. Physicists for the most part, are keen on the root or extreme reasons for material peculiarities and typically outline their understanding in numerical terms.

Physicists work across a wide scope of research fields, covering all length scales: from sub-nuclear and molecule physical science, through natural physical science, to cosmological length scales incorporating the universe in general (Oreskes, 2015). The field, for the most part, incorporates two kinds of physicists: theoretical physicists who represent considerable authority in the perception of normal peculiarities and the examination of investigations, and mathematical physicists who have some expertise in numerical calculation of physical structures to justify, clarify, and foresee regular phenomena. Physicists can provide their insight towards handling useful issues or to escalating new technologies (otherwise called applied physical science or engineering physics).

Pinwheel Galaxy

Pinwheel Galaxy is a spiral galaxy around 21 million light-years from Earth. Researchers call this whirling galactic system M101. You can view this heavenly body in the Ursa Major constellation, or “Big Dipper” in the Northern Hemisphere. On a dark, night sky you can view it with optics or a tiny telescope.

Pinwheel Galaxy (otherwise called Messier 101, M101 or NGC 5457) is a straight spiral galaxy around 21 million light-years from Earth in the star group Ursa Major. It was found by Pierre Méchain in 1781, and was communicated the same to the official astronomy catalog (Figure 141). On February 28th, 2006, NASA as well as the European Space Agency (ESA) provided a clear picture of the Pinwheel Galaxy, which was the biggest and comprehensive image of a universe by Hubble Space Telescope during the time. The picture was made out of 51 individual captions, in addition to some additional ground-based photographs (Oreskes, 2013).

Figure 141. Pinwheel galaxy has spiral arms like the Milky Way.

Source: https://theplanets.org/pinwheel/.

M101 is a huge galaxy, with a breadth of 170,000 light-years. By correlation, the Milky Way has a breadth of 100,000 to 120,000 light-years. It has approximately a trillion stars with a plate mass of 100-billion Sun masses, alongside a small focal lump of around 3 billion Sun masses. M101 has a high number of hot young stars which are ideal for making hot superbubbles. M101 is asymmetrical because of the tidal influence from interactions with its subordinate universes. These gravitational effects pack interstellar hydrogen gas, which triggers solid star development activity in M101's twisting arms that can be recognized in ultraviolet (UV) pictures.

Plain

A plain is an expansive space of level land. Plains are among the most common landforms, or forms of land, on the Earth. They comprise more than 33% of the world's landmass. Plains can be found on every continent. A large number of plains, like the Great Plains, which stretch across North America, are prairies. A meadow is a location where grass is the primary form of vegetation. In North America, mild plains—those in regions with warm summers and cold winters—are frequently called grasslands. In regions with minimal downpour and snow, short grasses develop. In regions that get more downpour and snow, tall grasses can reach 1.5 meters (5 feet) high. Notwithstanding, most tallgrass grasslands have been furrowed under and are presently farmland or field (Odenwald, 2020).

The Great Plains have sustained a wide range of cultures for millennia. The plain Indians for instance are multiple dozen clans, such as Blackfoot from the Canadian region of Alberta; Arapaho from the U.S. province of Wyoming; and Kickapoo, from the Mexican territory of Coahuila.

Tropical plains are known as savannas. Savannas exist in regions that are warm consistently. They regularly have dispersed trees. Savannas, for example, the Serengeti plains stretch across quite a bit of central Africa. They are additionally found in South America, Southern North America, and Australia. Not all plains are meadows. A few, like Mexico's Tabasco Plain, are wooded. Forested plains have various kinds of trees, bushes, and other foliage.

Planet

A planet is an enormous terrestrial or gaseous body that spins in a circle around the Sun or around another star, and that is not emanating energy from internal atomic reactions. Notwithstanding the above description, a few researchers add extra attributes like size, e.g., the item ought to be more than around 1,000 km [600 miles] across, or somewhat bigger than the biggest known space rock, Ceres), (i) shape (it ought to be adequately huge to have been just barely gotten by its own gravity into a circle—i.e., approximately 700 km or 435 miles across, based upon its density, or (ii) mass (it should have a low mass to prevent transitory atomic combination). Traditionally, it applies to bodies in Earth's planetary group including Mercury, Venus, Earth, Mars, Jupiter, Saturn, Uranus, and Neptune. Pluto was recorded as a planet until 2006 when it was removed (Neeck and Volz, 2011).

Everybody realizes that Earth, Mars, and Jupiter are planets. Yet, both Pluto and Ceres were once viewed as planets until new disclosures set off logical discussion regarding how to best depict them—a discussion that proceeds right up until today. The latest meaning of a planet was embraced by the International Astronomical Union in 2006. It says a planet should have three things:

- It should circle a star (in our case, the Sun);
- It should be sufficiently large to have enough gravity to mold it into a circular shape; and

- It should be large enough that its gravity cleared up other objects of a comparative size close to its rotation around the Sun.

Plate Tectonics

Plate tectonics is a geological theory that clarifies how key landforms are made because of Earth's underground movements. For plate tectonics, Earth's external layer, or lithosphere—comprised of the external and upper mantle—is divided into enormous rocky plates. Plate tectonics is a research hypothesis that clarifies how massive landforms are made because of Earth's underground movements. The hypothesis, which surfaced during the 1960s, changed Earth sciences by clarifying numerous peculiarities, including mountain creation events, volcanoes, and seismic tremors (Figure 142).

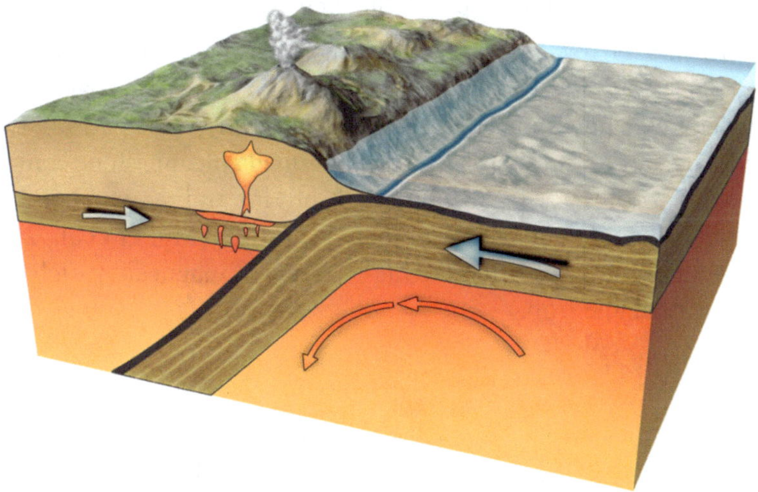

Figure 142. Plate tectonics causes the formation of new landmasses.

Source: https://www.collinsdictionary.com/dictionary/english/plate-tectonics.

In plate tectonics, the globe's outer layer, or lithosphere—comprised of the crust and upper mantle—is divided into enormous rocky plates. These plates rest on top of a partly molten rock layer known as the asthenosphere (Movva et al., 2012). Because of the convection of the lithosphere and asthenosphere, the plates move comparative with one another at various rates, from 2 to 15 centimeters each year. This relation of tectonic plates is liable for various topographical formations, for example, the Asian Himalaya Mountain range, the East African Rift, and San Andreas Fault in California, U.S.

The concept that continents shifted over the long run had been proposed before the 20th century. Be that as it may, a German researcher called Alfred Wegener transformed the scientific discussion. Wegener suggested that 200 million years prior, a supercontinent called Pangaea started to break into pieces, its parts getting away from each other. The continental landmasses we see today are sections of that supercontinent.

Plutonic

Plutonic is gotten from magma that has aired and hardened beneath the outer layer of the Earth. Plutonic igneous rocks were shaped by the sluggish hardening of magma far beneath the surface and defined by enormous precious stones. Named after Pluto, the Roman lord of the underworld. Examples incorporate granite, peridotite, and gabbro.

Gabbro is a coarse, dark tinted, intrusive molten stone that contains feldspar, pyroxene, and occasionally olivine. Granite is a coarse-grained, light-shaded, volcanic stone that contains primarily, feldspar, quartz, and mica minerals. Peridotite is a thick, coarse-grained volcanic stone comprising for the most part of the silicate compounds olivine and pyroxene. Peridotite is ultramafic, since it contains lower than 45% silica (Müller et al., 2019). It is rich in magnesium, mirroring the high extents of magnesium-rich olivine, containing calculable iron.

Polar

Polar region is found on the Northern and Southern Pole. It can also close, coming from, or identifying with either of the world's poles or the region inside the Arctic or Antarctic Circles. The northern polar location comprises predominantly of drifting and pack ice, 7–10 feet (2–3 m) dense, drifting on the Arctic Ocean and encompassed by landmasses. The southern polar ice cap averages 6,700 feet (around 2,000 m) in breadth, is underlain by the mainland landmass of Antarctica, and is encircled by seas (Figure 143).

The Arctic and Antarctica are enormous, frosty, and cold, however past that they are altogether diverse. Antarctica, in the south, is a mainland encircled by ocean; while the Arctic is an ocean encircled by landmasses. There are no local populations in Antarctica (yet lots of researchers); whereas the Arctic's local populace stretches back for a really long time. What is more, polar bears reside just in the Arctic; whereas penguins reside in Antarctica, and different sections of the Southern Hemisphere.

Figure 143. Polar bears living in the North Polar Region.

Source: https://www.britannica.com/animal/polar-bear.

Pollution

Pollution is the addition of unsafe or toxic compounds into the environment. These toxic compounds are known as pollutants. They can be natural, like volcanic debris. They can likewise be made from human action, for example, junk or spill over formed by processing plants. Pollutants harm the nature by air, water, and land. Numerous things that are helpful to individuals produce contamination (Metz, 2008). Vehicles cause contaminations from their exhaust pipes. Using coal to make power contaminates the air, etc. Businesses and homes produce trash and sewage that can contaminate the land and water. Pesticides—synthetic poisons used to exterminate weeds and bugs—saturate streams and harm natural life.

Every living thing—from single-celled organisms to blue whales—rely upon Earth's reserve of air and water. When these resources are dirtied, all types of life are compromised. Pollution is a worldwide issue. Although metropolitan regions are typically more contaminated than the open country, pollution can spread to remote regions where no individuals reside. For instance, pesticides, and different synthetics have been found in the northern Pacific Ocean, and Antarctic ice sheet. A tremendous assortment of small plastic particles known as the Great Pacific Garbage Patch.

Air and water flows convey pollution. Ocean flows and relocating fish transport marine pollutants all over. Winds can have radioactive material inadvertently let out of nuclear reactors and spread it all over the planet. Smoke from an industrial facility in one nation floats into another nation.

Porous

Porous stone is the fundamental component of hydrocarbon repositories. Oil and gas (or both) are produced from source layers, relocates upwards by dislodging water and is caught by overlying layers that will not permit hydrocarbons to move further upwards. Porous compounds in hydrocarbon reservoirs can be separated into carbonates and clastics. Clastics, for example, sandstone is made out of sandstone ordinarily stored in riverbeds throughout significant stretches of time and covered and packed over topographical periods. Carbonates (different calcium carbonate minerals) are normally created by organic cycles and again compacted by overlying material through significant stretches of time. Generally, 60% of regular oil and gas resources happen in clastics and 40% in carbonates (Li and Li, 2016).

Porous stone contains void space in which liquids, like compacted air, can be kept. Porosity is described as the level of a stone that is void and can be utilized for storage. A porosity of >10% is required for CAES (shale, sandstone, and limestone are examples of shakes). The ideal geographical design containing porous stone for CAES is a huge space of high porosity rock covered with impermeable stone. While porous stone is viewed as ideal for CAES, right now no CAES plants use it, and studies on utilizing them have been met with issues, for example, compacted air responding with pyrites in sandstone and low compressed wind current.

Precipitation

Precipitation is the formation of a solid in a liquid or within another solid during a synthetic process or by dissemination inside a solid. When the response happens in a fluid solution, the solid created is known as the precipitate. Precipitate is called a pellet following sedimentation when utilizing a rotator to squeeze it into a lesser mass. The precipitate-free fluid leftover over the solid is known as the supernate or supernatant. Powders gotten from precipitation are called blossoms.

Precipitates can develop when:

- Two dissolvable salts react in the solution to create at least one insoluble item;
- The solution temperature is reduced. The reduced temperature lessens the solvency of a salt, causing its precipitation to be solid. The substance that causes the solid to shape is known as the precipitant (Ledley et al., 2011);
- Utilizing a supersaturated solution. Without adequate gravitational effect (settling) to merge the solid particles, the precipitate stays in suspension;
- Anti-solvent is included. This profoundly drops the solvency of the ideal product. Hence, the precipitate may effortlessly be separated through filtration/emptying or centrifugation.

Precipitate improvement is important in the identification of the form of cation within a salt. For instance, a barium nitrate solution reacts with sulfate particles to create a solid barium sulfate precipitate, showing that it is normal that sulfate particles are available. Precipitation reactions are additionally helpful to remove components, such as magnesium, from seawater.

Precipitate properties include:

- Insoluble ionic solid compounds of a reaction;
- Responses reliant upon temperature or on solution fixation (Lewis, 2008);
- Solids created in precipitate reactions are translucent and can be suspended as silt;
- The two sections of supernatant fluid (precipitate and supernate) may easily be separated.

Psychrometer

The psychrometer is a hygrometer that uses two thermometers—a wet and dry-bulb—to determine moisture levels through evaporation. The distinction in temperature between wet and dry bulb thermometers can be utilized to process the level of water vapor noticeable in the air (Figure 144).

The instrument, known as psychrometer, was later manufactured by Rudolf Fuess' in Germany. The legendary Assmann-type psychrometer was made out of two indistinguishable mercury-in-synthetic thermometers, ventilated through a mechanical fan, and providing reliable moisture estimations (Košler, 2007). Be that as it may, constant RH assessment was unrealistic and any examination of the dry and wet bulb temperatures needed the support of tables or mechanical structures to be deciphered. Consequently, the mechanical psychrometer was seldom utilized for protection purposes and was bound to research center alignment of lower exactness instruments, for example the hair hygrometer.

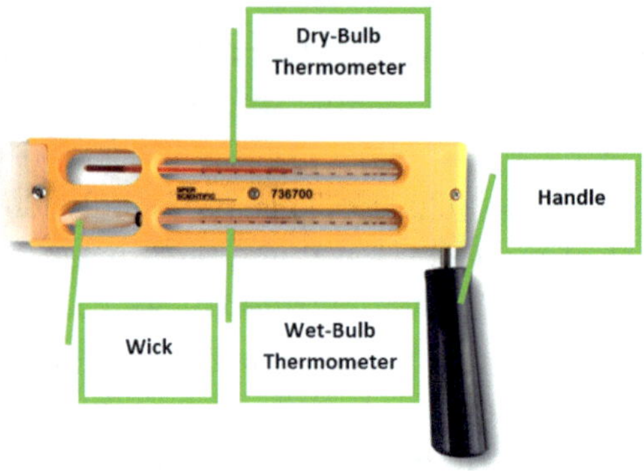

Figure 144. Psychrometer is a device used to measure humidity.

Source: https://www.instrumentchoice.com.au/news/how-does-a-sling-psychrometer-work.

Today, the notable Assmann psychrometer has been deserted and has been subbed with mechanical electric fans and further developed temperature sensors, for example platinum obstruction, thermocouples, and thermistors. Electric psychrometers are suggested by EN16242: 2012 to intermittently control if other moisture sensors (for example slight film capacitive or resistive sensors) give precise readings or need alignment. Nonetheless, it tends to be likewise associated with a data lumberjack and with a water repository that permits medium-term impartiality, it might give routine observation at chosen inspection stretches.

The psychrometer is a tool dependent on the readings of a matched set of sensitive thermometers (for example they should read the same at some random temperature), joined one next to the other: one being typical, called the 'dry bulb' to be recognized from the other that has its bulb covered with a wet cotton material wick, called 'wet bulb' (Kleinhans et al., 2010). At equilibrium, for example, when the thermal Qυ lost by evaporation from the wet bulb rises to the reasonable heat Qs moved from the surrounding air to the cooler wet bulb.

R

Radar

Radar (radio detection and ranging) is a detecting technology that employs radio waves to calculate object distance, angle, and velocity. Aircraft, ships, satellites, guided missiles, motor vehicles, weather formations, and topography can all be detected with it. A radar system consists of a transmitter that generates electromagnetic waves in the radio or microwave frequency range, a sending antenna, a receiving antenna (typically the same antenna), and a receiver and processor that determines the object's attributes (s). The transmitter's radio waves (pulsed or continuous) reflect off the object and return to the receiver, providing data on the item's location and speed (Figure 145).

Figure 145. Long-range radar antenna, used to track space objects and ballistic missiles.

Source: https://en.wikipedia.org/wiki/Radar#/media/File:Radar_antenna.jpg.

Several countries developed radar in secret for military use in the years leading up to and during World War II. The cavity magnetron, developed in the United Kingdom, was a crucial development that permitted the production of relatively tiny devices with sub-meter resolution.

The United States Navy originated the term RADAR in 1940 as an acronym for "radio detection and ranging." Since then, the term radar has become a common noun in English and other languages, shedding all capitalization. "Radio azimuth direction and range" was suggested during RAF RADAR courses at Yatesbury Training Camp in 1954/5 (Klippel et al., 2019).

Radiation

Radiation is defined as the emission or transmission of energy in the form of waves or particles over space or a material medium in physics. This includes the following:

- Particle radiation, such as alpha radiation (), beta radiation (), proton radiation (), and neutron radiation () electromagnetic radiation, such as radio waves, microwaves, infrared, visible light, ultraviolet (UV), x-rays, and gamma radiation () particle radiation, such as alpha radiation (), beta radiation (), proton radiation, and neutron radiation (particles of non-zero rest energy).
- Ultrasound, sound, and seismic waves are examples of acoustic radiation (dependent on a physical transmission medium).
- Gravity radiation, which manifests as gravitational waves or ripples in spacetime curvature.

Depending on the energy of the radiated particles, radiation is frequently classified as ionizing or non-ionizing. Ionizing radiation has a wavelength of more than 10 eV, which is sufficient to ionize atoms and molecules as well as disrupt chemical bonds (Hundman and Mattmann, 2017). Because of the significant disparity in harmfulness to living organisms, this is a crucial distinction.

Radiator

Radiators are heat rejection systems that use radiating surfaces with strong infrared emissivity and low solar absorptance to reject waste thermal energy from a heat source to outer space. Radiators are divided into passive and active categories based on how heat is transported from the source to the radiating surfaces.

Radio Wave

Radio waves are a type of electromagnetic radiation. The wavelength of a radio wave is substantially longer than that of visible light. Radio waves are widely used by humans for communication.

Radio waves have wavelengths ranging from a few millimeters (tenths of an inch) to hundreds of kilometers (hundreds of miles). For instance, visible light has wavelengths between 400 to 700 nanometers, which is around 5,000 times shorter than the shortest wavelength radio waves. Radio waves have frequencies ranging from a few kilohertz (kHz) to a few terahertz (thousands of hertz) (THz or 1,012 hertz). Along the electromagnetic spectrum, "far infrared" radiation borders radio waves and has somewhat more energy and shorter wavelengths than radio waves (Guo et al., 2016).

Microwaves are radio waves with a short wavelength that we utilize for cooking and communication. The wavelengths of microwaves range from a few millimeters to tens of centimeters (tenths of inches to tens of inches).

Television and FM and AM radio broadcasts, military communications, mobile phones, ham radio, wireless computer networks, and a variety of other communications applications all employ different frequencies of radio waves (Figure 146).

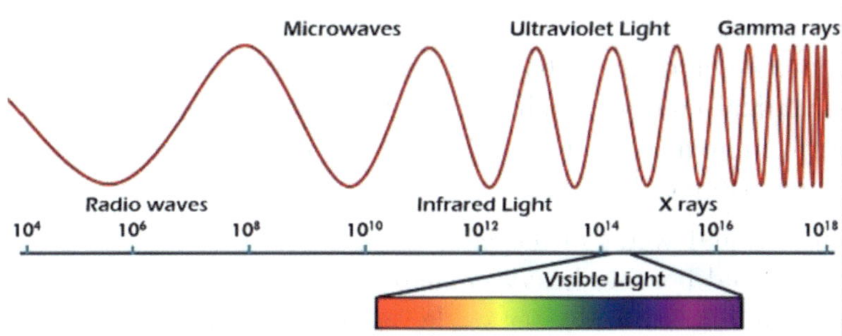

Figure 146. Different types of waves.

Source: https://techdifferences.com/difference-between-radio-wave-and-microwave.html.

The majority of radio waves are unaffected by the Earth's atmosphere. Some frequencies, on the other hand, can be reflected or absorbed by charged particles in the ionosphere.

Radioactive Decay

The process by which an unstable atomic nucleus loses energy through radiation is known as radioactive decay (also known as nuclear decay, radioactivity, radioactive disintegration, or nuclear disintegration). The term "radioactive" refers to a substance that contains unstable nuclei. Alpha decay (-decay), beta decay (-decay), and gamma decay (-decay) are three of the most prevalent types of decay, all of which entail the emission of one or more particles. The weak force is responsible for beta decay, whereas the electromagnetic and strong forces are in charge of the other two.

At the atomic level, radioactive decay is a stochastic (i.e., random) process. According to quantum theory, no matter how long an atom has lived, it is impossible to forecast when it will decay. The total decay rate, on the other hand, can be stated as a decay constant or as a half-life for a large number of similar atoms (Goldstein et al., 2017; Escorihuela, 2018).

Rainfall

Rain is liquid water in the form of droplets that have condensed from atmospheric water vapor and become heavy enough to fall. Rain is an important part of the water cycle since it is responsible for depositing the majority of the world's fresh water. It provides habitat for a variety of habitats as well as water for hydroelectric power facilities and farmland irrigation.

Moisture moving along three-dimensional zones of temperature and moisture contrasts known as weather fronts is the primary cause of rain production. Precipitation falls from convective clouds (those with strong upward vertical motion), such as cumulonimbus (thunder clouds), which can organize into narrow rain bands if enough moisture and upward motion are sufficient (Figure 147).

Rain is liquid water in the form of droplets that condensed from air water vapor and became heavy enough to fall due to gravity. Rain plays an important role in the water cycle, as it is responsible for depositing the majority of the world's fresh water. It provides ideal conditions for a variety of ecosystems, as well as water for hydropower plants and farmland irrigation.

Moisture traveling over three-dimensional zones of temperature and moisture differences called weather fronts is the primary driver of rain formation. Convective clouds (those with strong upward vertical motion) such as cumulonimbus (thunder clouds) can form narrow rain bands if enough moisture and upward motion are present (Duffy, 2014). Intense rainfall with a storm is called a rainstorm.

Figure 147. Rain alert in Maharashtra.

Source: https://www.india.com/news/delhi/delhi-experiences-second-highest-annual-rainfall-in-121-years-since-1933-1502-8mm-rainfall-recorded-5069324/.

Rate

A certain quantity or amount of one thing considered in relation to a unit of another thing and used as a standard or measure. Earth rate is the angular velocity or rate of the earth's rotation.

Reaction

When two or more molecules contact and change, it is called a reaction. To generate new molecules, atoms' bonds are disrupted and new ones are formed.

A chemical reaction occurs when one or more compounds, known as reactants, are changed into one or more distinct substances, known as products. Chemical elements or compounds are examples of substances. The constituent atoms of the reactants are rearranged in a chemical reaction, resulting in the formation of various substances as products.

A physical response is one that causes a change in a materials or substance's physical qualities. Density, mass, and volume are examples of physical qualities. The definition of a physical reaction is one in which molecules undergo molecular rearrangement but are not chemically changed.

Reactor

- A person who reacts.
- A device for introducing reactance into an alternating-current circuit (such as a coil, winding, or low-resistance conductor) (Clowes, 2010).
- A device for the controlled release of nuclear energy or a vat for an industrial chemical reaction (as for producing heat).

Recharge

A hydrologic process in which water travels downhill from surface water to groundwater is known as groundwater recharge, deep drainage, or deep percolation. Recharge is the most common way for water to enter an aquifer. This process normally happens below plant roots in the vadose zone and is often described as a flux to the water table surface. Water moving out from the water table and into the saturated zone is also referred to as groundwater recharge. Rainwater and or recovered water are channeled to the subsurface in both natural (via the water cycle) and anthropogenic (i.e., "artificial groundwater recharge") processes (Figure 148).

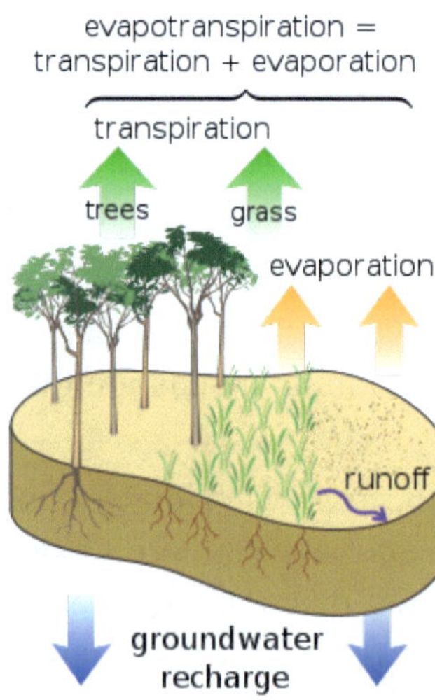

Figure 148. Water recharge.

Source: https://en.wikipedia.org/wiki/File:Surface_water_cycle.svg.

Rain and snowmelt replenish groundwater, and surface water replenishes it to a lesser level (rivers and lakes). Human activities like as paving, development, and logging may obstruct recharge in several ways. These operations can cause topsoil loss, which can lead to decreased water infiltration, increased surface runoff, and decreased recharge. Groundwater use, particularly for irrigation, has the potential to deplete water levels (Clary and Wandersee, 2011). Because the volume-rate extracted from an aquifer in the long run should be less than or equal to the volume-rate replenished, groundwater recharge is an important process for sustainable groundwater management.

Recharge can assist in the movement of surplus salts from the root zone to deeper soil layers or into the groundwater system. Water saturation in groundwater is increased by tree roots, which reduces discharge. Flooding promotes aquifer recharge by temporarily increasing river bed permeability by pushing clay soils downstream.

In India, where over-pumping of groundwater by farmers has depleted subterranean resources, artificial groundwater recharge is becoming increasingly significant. In 2007, the Indian government allocated 1,800 crore (equivalent to 46 billion or US $610 million in 2020) to fund dug-well recharge projects in 100 districts across seven states where water stored in hard-rock aquifers had been over-exploited, based on recommendations from the International Water Management Institute (Cossu et al., 2010). Waste disposal through water flux, such as dairy farms, industrial, and urban runoff, is another environmental challenge.

Reflect

The rapid change in the propagation direction of a wave that reaches the boundary between two mediums is known as reflection. At least a portion of the approaching wave disturbance is contained within the same medium. At plane boundaries, regular reflection occurs according to a simple law. The angle between the approaching wave's direction of motion and a perpendicular to the reflecting surface (angle of incidence) is equal to the angle between the reflected wave's direction of motion and a perpendicular to the reflecting surface (angle of reflection). Diffuse reflection occurs at uneven or irregular boundaries. The fraction of an oncoming wave's energy reflected by a surface material is called reflectivity (Figure 149).

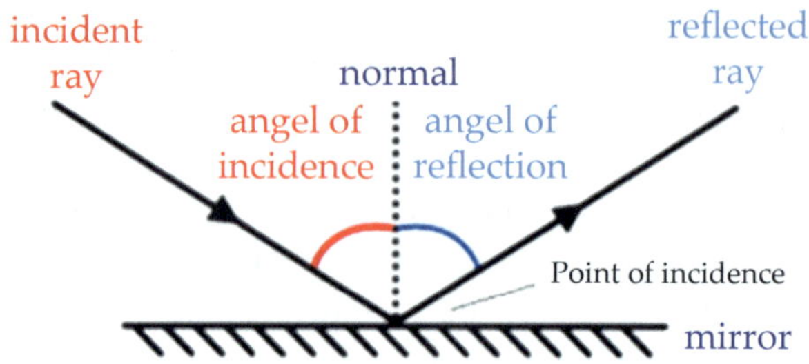

Figure 149. Reflection of light.

Note: In the reflection of light, the angle of incidence is equal to the angle of reflection, measured from the normal (the line perpendicular to the point of impact).

Source: https://thefactfactor.com/facts/pure_science/physics/reflection-of-light/10809/.

Region

The geological study of large-scale regions is known as regional geology. It usually entails a combination of geological disciplines in order to piece together a region's past. It is geology's version of regional geography. Geologically significant limits and the presence of geologic processes define the size and borders of each region. The interfingering facies change in sedimentary deposits when discussing a sedimentary basin system, or the leading or boundary thrust of an orogen are examples of geologically significant boundaries (Cerase, 2019).

Relative Humidity

The amount of water vapor in the air is referred to as humidity. The humidity will be high if there is a lot of water vapor in the air. The wetter it feels outside as the humidity rises.

Humidity is frequently described as relative humidity in weather reports. The amount of water vapor in the air, represented as a percentage of the maximum amount of water vapor the air can hold at the same temperature, is known as relative humidity. Consider the air at a freezing –10°C (14°F). The air can only retain 2.2 grams of water per cubic meter at such temperature. When it is –10°C outside and there is 2.2 grams of water per cubic meter, we are at an uncomfortably high 100% relative humidity. We are at 50% relative humidity if there is 1.1 grams of water in the air at –10°C (Figure 150).

Figure 150. Ekom, Cameroon, is very hot and humid. Near this waterfall, the humidity, or amount of water vapor suspended in the air, is so high that it is not easy for people to breathe.

Source: https://www.nationalgeographic.org/encyclopedia/humidity/.

Relativity

The term "relativity" refers to how things are only significant in connection to other things. In physics, a hypothesis proposed by Albert Einstein that all motion must be characterized in terms of a frame of reference, and that space and time are relative rather than absolute concepts: it is divided into two sections (Britner, 2008). The theory of uniform motion (special theory of relativity, or special relativity) is based on the two postulates that physical laws have the same mathematical form in any inertial system and that the velocity of light is independent of the motion of its source and will have the same value when measured by observers moving at constant velocity with respect to each other.

The inferences that can be drawn from these postulates are that there can be no motion faster than the speed of light in a vacuum, mass grows as velocity increases, mass, and energy are equal, and time is dependent on the relative motion of the observer measuring the time. The theory of gravity (general theory of relativity, or general relativity) is based on the assumption that the local effects of a gravitational field and an inertial system's acceleration are the same.

Resource

The term "resource" refers to all of the items in our environment that assist us meet our needs and desires. Resources are divided into two categories based on their availability: renewable and non-renewable resources. They can also be characterized as actual or potential depending on their level of development and use, as biotic or abiotic depending on their origin, and as ubiquitous or localized depending on their distribution (private, community-owned, national, and international resources). With the passage of time and the advancement of technology, an object becomes a resource (Butler, 2016). Increased wealth, effective system functioning, or improved well-being are all possible outcomes of resource consumption. A natural resource is anything that is collected from the environment to satisfy human needs and desires. A resource meets the demands of a live organism in a broader biological or ecological sense.

The notion of resources has been created in a variety of fields, including economics, biology, and ecology, computer science, management, and human resources, and is linked to concepts such as competition, sustainability, conservation, and stewardship. Commercial and non-commercial aspects in human civilization necessitate resource allocation through resource management.

Rock Salt

Rock salt refers to a sedimentary rock that is almost entirely made of halite, a sodium chloride (NaCl) mineral. It arises when substantial amounts of seawater or salty lake water evaporate from an arid-climate region with a renewing flow of salt water but little intake of other water.

In many places of the world, rock salt deposits occur as laterally widespread rock formations. Except in arid locations, these water-soluble pebbles are rarely found on the Earth's surface. The United States' most geographically vast salt layers cover hundreds of square miles and have a thickness of over 100 feet (Bohle, 2015).

For thousands of years, people have mined rock salt or created salt through evaporation. The majority of ancient salt was utilized in cooking and food preservation.

Rod

A conductor on top of a building to protect the building in the event of lightning by taking the charge harmlessly to earth. A metal rod can also be embedded in the ground to make a ground connection to the earth.

Control rods are used to control the rate of fission in a nuclear reactor. Their compositions include chemical elements such as boron, cadmium, silver, hafnium, or indium, that are capable of absorbing many neutrons without themselves fissioning.

Rotation

The strict definition of rotation is "the circular movement of an object about a point in space." This is used in geometry as well as astronomy and physics.

An object rotates when it moves in a circular motion around its center of rotation. A rotation axis is an imaginary line around which three-dimensional objects such as the earth, moon, and other planets always revolve. When the axis passes through the center of mass of the body, the body is said to rotate or spin (Bohle, 2015). Imagine a line passing through the center of Earth that goes through both the North Pole and the South Pole. This imaginary line is called an axis. Earth spins around its axis, just as a top spin around its spindle. This spinning movement is called Earth's rotation (Figure 151).

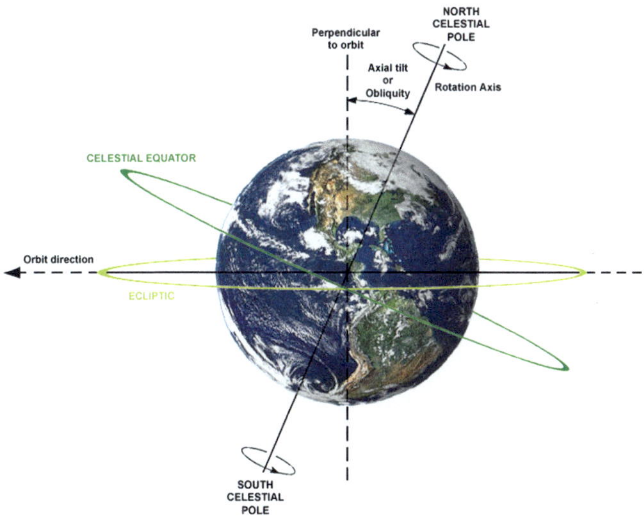

Figure 151. Earth's rotation.

Source: https://courses.lumenlearning.com/geophysical/chapter/earths-motions/.

Rupture

To breach or disturb (a harmonious situation). Surface rupture (also known as ground rupture or ground displacement) is the visible offset of the ground surface caused by an earthquake rupture along a fault. Buried rupture, in which there is no displacement at ground level, opposes surface rupture. In addition to the potential of Earth shaking, any structure placed across an active fault zone faces a significant risk. Surface rupture refers to movement on either side of a ruptured fault, which might be vertical or horizontal. Large tracts of land can be affected by surface rupture (Zhuang et al., 2019).

S

Saturation

Saturation is a physical or chemical situation characterized by the existence of an equilibrium between opposing forces or an exact balance of opposing process speeds.

Saturation in the Earth sciences refers to the amount of water in the soil, with the unsaturated zone above the water table and the saturated zone below. Without boiling or condensing, a saturated liquid or saturated vapor holds as much thermal energy as it can.

Saturation happens when water fills the microscopic pores and crevices in rock and soil beneath the earth's surface, which occurs below the water level. Saturated soil is less corrosive than material above the saturation point.

Scale

Scale (chemistry), the range of mass or volume of a chemical reaction or process.

Sea Floor

The ocean floor, often known as the sea floor, is the ocean's actual floor. When you dive, you are at the bottom of the ocean. Few people have ventured to the ocean's deepest depths, as it necessitates the use of specialized diving equipment due to the extremely high-water pressure as you descend. When you are that far down, you cannot afford to be without protective gear. The water's pressure would crush you (Figure 152).

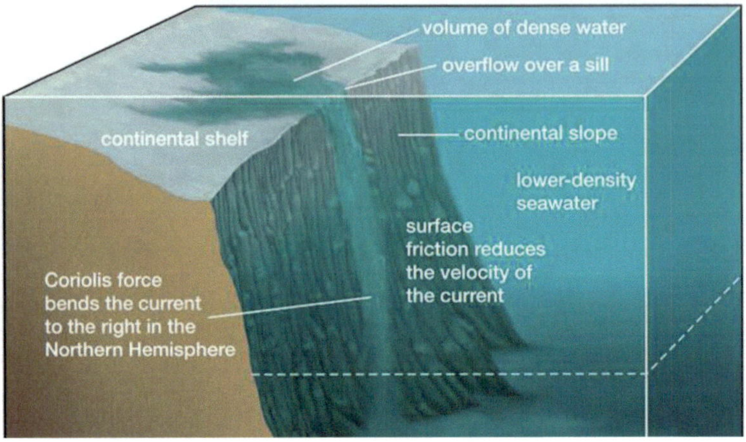

Figure 152. Ocean floor.

Source: https://www.britannica.com/science/ocean-floor.

Other techniques, such as radar, are being used by scientists to investigate the ocean floor. The ocean floor was mapped to a resolution of 1.5 kilometers in 2015, according to Scientific American. This means that anything greater than 1.5 kilometers (one mile) will be visible. The canyons, abysses, and slopes of the ocean floor would be visible thanks to this mapping.

Sea Level

The base level for measuring height and depth on Earth is the sea level. The ocean's surface tends to seek the same level throughout the world because it is one continuous body of water. However, fluctuations in gravity and temperature, as well as winds, currents, and river discharges, prevent the sea surface from being truly level.

The notion of "local mean sea level" was devised so that the ocean's surface could be used as a base for measuring altitudes. Local mean sea level in the United States and its territories is calculated by obtaining hourly sea level measurements at several places over a 19-year period and then averaging the results.

A Metonic cycle is defined as a 19-year span. It allows scientists to account for the moon's long-term orbital fluctuations. It also takes into account the effects of weather and oceanic conditions in the area (Yu and Gong, 2012).

The sea level is measured in relation to the land surrounding it. The elevation of land, like the water, can rise and fall throughout time. The immense weight of a glacier on land, for example, pushes the land down, bringing it closer to sea level. If the ice retreats, or melts, and its weight is removed, the same land bounces back (a process known as post-glacial rebound). Measurements of local mean sea level are a combination of sea level changes and land mobility.

Sediment

Solid material that has been transferred and deposited in a new location is known as sediment. Sediment can be made up of rocks and minerals, as well as plant and animal remain. It can range in size from a grain of sand to a boulder. Erosion is the process by which sediment moves from one location to another. Erosion is the process of removing and transporting rock or soil. Sediment can be moved by erosive forces such as water, ice, or wind (Witze, 2013).

Sediment, such as gravel or pebbles, can be washed down a creek, into a river, and eventually to the delta of that river. Sediment accumulates in many places, including deltas, river banks, and the bottoms of waterfalls.

As the ice carves its way through the landscape or melts, glaciers can freeze sediment and deposit it elsewhere. Moraine is a type of sediment generated and deposited by glaciers (Figure 153).

Figure 153. Sediment forms a spiral pattern as it flows from the Port River mouth in Adelaide, Australia. Sediment can consist of rocks and minerals, as well as the remains of plants and animals.

Source: https://www.nationalgeographic.org/encyclopedia/sediment/.

Dust storms or sandstorms can blow dirt across a plain. Sand dunes are formed up of stony material that has been eroded down by wind and sand particles colliding with each other.

Sediment is significant because it frequently adds nutrients to the soil. Sediment-rich areas are frequently also rich in biodiversity. Sedimentary soil is usually preferable for agriculture. Deltas and river banks, which produce a lot of silt, are frequently the most fertile agricultural sites in a region.

The Nile River flooded every year for thousands of years, bringing with it 4 million metric tons (4.4 million short tons) of nutrient-rich material. The Nile's banks are still Egypt's most fertile agricultural region (Wilkinson et al., 2020).

Sedimentary Rock

Sedimentary rocks are rocks that form at or near the Earth's surface as a result of the lithification of sediment (detrital rock) or the precipitation of solution at typical surface temperatures (chemical rock). Sedimentary rocks are the most common rocks exposed on the Earth's surface, yet they make up just a small part of the crust, which is dominated by igneous and metamorphic rocks.

The buildup and lithification of sediment (detrital rock) or precipitation from solution at typical surface temperatures generate sedimentary rock at or near the Earth's surface (chemical rock). Sedimentary rocks are the most common rocks exposed on the Earth's surface, yet they make up just a small part of the crust, which is dominated by igneous and metamorphic rocks (Wallace-Wells, 2019).

Layers of sediment may accumulate and form into sedimentary rock over millions of years. Sandstone, rock salt, and coal are all examples of sedimentary rock.

Sandstone is formed when sand hardens. Sandstone has been blended with sticky cement to make concrete for ages. Concrete is an important building material that is utilized in the construction of numerous structures and roadways.

As oceans evaporate, rock salt, also known as halite, develops. The oceans are composed of salt water. When water evaporates into the atmosphere, it leaves the salt behind. The Bonneville Salt Flats are flat desert lands in the U.S. state of Utah that are covered with a coating of rock salt sediment. The ancient sea that originally covered the area, Lake Bonneville, has long since vanished.

Coal is a sediment composed of hardened plant debris. Coal, which may be found on every continent except Antarctica, can be found on the locations of old marshes.

Seismic

Relating to an earthquake or to other tremors of the Earth, such as those caused by large explosions: a seismic disturbance.

Seismic waves are elastic waves that travel through solid or fluid media. They are classified as body waves, which travel through the materials' interiors, surface waves, which travel over surfaces or interfaces between materials, and normal modes, which are a type of standing wave.

1. **Body Waves:** These are classified into two types: pressure waves (P-waves) and shear waves (S-waves) (S-waves). P-waves are longitudinal waves that entail compression and expansion along the wave's direction of motion. They are always the first waves to emerge on a seismogram since they are the quickest traveling waves through solids (Toth et al., 2014). S-waves are transverse waves that travel perpendicular to the propagation direction. S-waves move more slowly than P-waves. As a result, they appear later on a seismogram than P-waves. Because of their low shear strength, fluids cannot support transverse elastic waves, hence S-waves can only move in solids.
2. **Surface Waves:** These are caused by P- and S-waves interacting with the Earth's surface. Because these waves are dispersive, various frequencies have varying velocities. Rayleigh waves, which have both compressional and shear motions, and Love waves, which are solely shear, are the two basic forms of surface waves. Rayleigh waves form when P-waves and vertically polarized S-waves contact with a solid surface and can exist in any solid medium. Love waves are created by horizontally polarized S-waves interacting with the surface and can exist only if the elastic characteristics of a solid material change with depth, which is always the case in seismological applications. Surface waves travel at a slower rate than P- and S-waves because they are the result of these waves taking indirect paths to interact with the Earth's surface.

3. **Normal Modes:** Although both body and surface waves are moving waves, major earthquakes can cause the entire Earth to "ring" like a resonant bell. This ringing is a combination of typical modes with discrete frequencies and periods of an hour or less. A major earthquake's normal mode motion can be noticed for up to a month following the event. The first observations of normal modes were made in the 1960s, when the introduction of greater fidelity instrumentation coincided with two of the 20th century's greatest earthquakes, the 1960 Valdivia earthquake, and the 1964 Alaska earthquake. Since then, the Earth's normal modes have provided some of the most powerful restrictions on the deep structure of the Earth (Tollefson, 2015).

Semiarid

A semi-arid climate, also known as a semi-desert climate or steppe climate, is one in which precipitation is less than potential evapotranspiration but not as low as a desert climate. Semi-arid climates come in a variety of forms, depending on factors such as temperature, and they give rise to several biomes; characterized by light rainfall especially: having from about 10 to 20 inches (25 to 51 centimeters) of annual precipitation.

The precipitation requirement must first be determined to determine if an area has a semi-arid climate. The following approach was used to calculate the precipitation threshold (in millimeters):

- If at least 70% of total precipitation falls in the high-sun part of the year (April–September in the northern hemisphere, October–March in the southern hemisphere), multiply the average annual temperature in degrees Celsius by 20 and add 280.
- If 30–70% of total precipitation falls in the high-sun half of the year, multiply by 140.
- If less than 30% of total precipitation falls in the high-sun half of the year, add nothing.
- If the annual precipitation in millimeters is less than the threshold but greater than half the threshold, the area is designated as a BS (steppe or semi-arid climate).

Furthermore, three widely used isotherms are used to distinguish hot semi-arid climates from cold semi-arid climates: a mean yearly temperature of 18°C (64.4°F), a mean temperature of 0°C (32.0°F) in the coldest month, or a mean temperature of 3°C (26.6°F) in the coldest month (Thessen,

2016). A site with a BS-type climate is classed as hot semi-arid (BSh) if its relevant mean temperature is higher than the specified isotherm, whereas a location with a temperature lower than the given isotherm is categorized as cold semi-arid (BSc) (BSk).

Sensor

In contrast to in situ or on-site observation, sensing is the acquisition of information about an object or phenomenon without having direct contact with it. The phrase is used to describe the process of gathering knowledge about the Earth and other planets. Remote sensing is used in a variety of fields, including geography, land surveying, and most Earth science disciplines (for example, hydrology, ecology, meteorology, oceanography, glaciology, and geology), as well as military, intelligence, commercial, economic, planning, and humanitarian applications.

The phrase "remote sensing" now refers to the detection and classification of objects on Earth using satellite or aircraft-based sensor technologies. Based on propagating signals, it includes the surface, atmosphere, and seas (e.g., electromagnetic radiation). It can be divided into "active" remote sensing (when a signal is emitted by a satellite or aircraft to the object and its reflection is detected by the sensor) and "passive" remote sensing (when a signal is emitted by a satellite or aircraft to the object and its reflection is detected by the sensor).

Shoreline

The land near the edge of a big body of water, such as an ocean, sea, or lake, is known as a shore or coastline. A shore is the larger fringe that is geologically transformed by the action of the body of water past and present in physical oceanography, whereas the beach is at the edge of the shore, indicating the intertidal zone where one exists. A shore, unlike a coast, can be located on any body of water, whereas a coast must be located on an ocean or sea. As a result, a coast is a type of shore in that meaning. However, the term "coast" frequently refers to an area that extends much beyond the shoreline, typically for miles into the interior (Telling et al., 2017).

Shorelines are influenced by the surrounding landscape's topography as well as water-induced erosion, such as waves. The geological composition of rock and soil dictates the type of shore which is created.

Silica

Silica, commonly known as silicon dioxide, is a compound of silicon and oxygen, the two most abundant elements in the Earth's crust, SiO_2. The major element of more than 95% of known rocks is silica, which makes up 59% of the Earth's crust. Quartz (by far the most common), tridymite, and cristobalite are the three principal crystalline types of silica. Coesite, keatite, and lechatelierite are some of the other variations. Silica sand is found in sandstone, Portland cement, concrete, and mortar, as well as in structures and roadways. Silica is also used to grind and polish glass and stone, as well as in foundry molds, the production of glass, ceramics, silicon carbide, ferrosilicon, and silicones, as a refractory material, and as a gemstone. Desiccants such as silica gel are frequently used to eliminate moisture (Figure 154).

Figure 154. Silica gel.

Source: https://www.britannica.com/science/silica-gel.

Silicate

Silica, commonly known as silicon dioxide, is a compound of silicon and oxygen, the two most abundant elements in the Earth's crust. The major element of more than 95% of known rocks is silica, which makes about 59%

of Earth's crust. Quartz (by far the most common), tridymite, and cristobalite are the three principal crystalline types of silica. Coesite, keatite, and lechatelierite are some of the others (Showstack, 2008). Portland cement, concrete, and mortar, as well as sandstone, contain silica sand, which is utilized in the construction of buildings and roads. Silica is also used in the manufacturing of glass, ceramics, silicon carbide, ferrosilicon, and silicones, as a refractory material, and as gemstones. Desiccants like silica gel are commonly used to absorb moisture.

Silt

Silt is a solid, dust-like silt that is transported and deposited by water, ice, and wind. Silt is made up of larger than clay but smaller than sand rock and mineral particles. Silt particles are so tiny that they are difficult to see individually. A particle must be less than.005 centimeters (.002 inches) in diameter to be categorized as silt. Silt, along with other forms of sediment such as clay, sand, and gravel, can be found in soil.

When wet, silty soil is slick, not gritty, or rough. If the silt component of the soil is greater than 80%, it is classified as silt. Siltstone is formed when silt deposits are squeezed and the grains are pushed together (Schmitz, 2011).

Silt is formed when water and ice erode (or wear away) rock. As tiny rock fragments are carried by rushing water, they scrape against the sides and bottoms of stream beds, chipping away at the rock. The particles grind against each other, reducing in size until they are silt-sized. Glaciers can also erode rock particles, resulting in the formation of silt. Finally, wind can blow rock particles down a canyon or over a landscape, causing them to grind against the canyon wall or one other. All three processes result in the formation of silt.

Silt has the ability to alter landscapes. Silt settles in stagnant water, for example. As a result, silt deposits gradually fill in wetlands, lakes, and harbors. Silt is deposited along river banks and flood plains during floods. Rivers deposit silt as they drain into another body of water, forming deltas. Silt makes up over 60% of the Mississippi River Delta.

Sink

A sink is a reservoir that stores a substance in the context of a biogeochemical cycle. Plants, for example, extract carbon dioxide (CO_2) from the atmosphere through photosynthesis, and the carbohydrates they generate serve as a storage reservoir. As a result, plant matter acts as a carbon sink in the carbon cycle (Schmitz, 2012).

Skeleton

A skeleton is the structural framework that holds an animal's body together. There are three sorts of skeletal structures: the exoskeleton, which is an organism's stable outer shell; the endoskeleton, which provides support inside the body; and the hydroskeleton, which is a flexible skeleton supported by fluid pressure.

The hydrostatic skeleton, exoskeleton, and endoskeleton are three different skeleton architectures that give these services to organisms.

Slope

When building or planting, the slope of the soil is a crucial factor to consider. The slope gradient is the percent rise or fall of the soil surface from horizontal over a distance of 100 feet, given as an angle of inclination or drop.

The flow of water that can erode the soil is influenced by the slope of the soil. It also has an impact on the use of machinery, building construction, landscaping, maintenance, and on-site septic tank waste disposal systems (because seepage can occur down-slope of an absorption field).

Different activities are better suited to different slope grades (see sidebar). A "gently sloping" slope of 2–6% is suggested for residential construction, for example. Water does not drain away from the home if the slope is too level. Erosion and soil stability might be an issue if the slope is too steep (Rao, 2012).

Snowfall

Individual ice crystals grow while suspended in the atmosphere—usually within clouds—and then descend to the earth, where they undergo further transformations. Throughout its life cycle, it is made up of frozen crystalline water, beginning when ice crystals form in the atmosphere, grow to millimeter size, precipitate, and pile on surfaces, metamorphosis in place, and eventually melt, slide, or sublimate away.

Snowstorms feed on sources of atmospheric moisture and cold air to form and grow. Snowflakes form when supercooled water droplets collide with particles in the atmosphere, forming hexagonal-shaped crystals. Platelets, needles, columns, and rime are some of the most common shapes found in snowflakes (Rivet and Kastens, 2012). Snow may blow into drifts as it accumulates into a snowpack. Sintering, sublimation, and freeze-thaw transform accumulated snow over time. A glacier can form where the environment is cold enough for year-to-year accumulation.

Snow, on the other hand, melts seasonally, causing runoff into streams and rivers and replenishing groundwater.

The polar regions, the northernmost half of the Northern Hemisphere, and mountainous places globally with sufficient moisture and low temperatures are all major snow-prone zones. Apart from Antarctica, snow is confined to mountainous places in the Southern Hemisphere.

Transportation, agriculture, sports including skiing, snowboarding, and snowmachine travel, and warfare are all affected by snow. Snow has an impact on ecosystems by providing an insulating layer in the winter that allows plants and animals to endure the cold.

Soil

Soil is a natural body made up of solids (minerals and organic matter), liquids, and gases that occurs on the land surface, occupies space, and is distinguished from the initial material by one or both of the following: horizons, or layers, that are distinguishable from the initial material as a result of additions, losses, transfers, and transformations of energy and matter or the ability to support rooted plants in a natural environment.

The barrier between soil and air, shallow water, live plants, or plant materials that have not begun to disintegrate is the top limit of soil. If the surface is continually covered by water that is too deep (usually more than 2.5 meters) for the growth of rooted plants, the area is not regarded to have soil.

The lower boundary between soil and nonsoil beneath it is the most difficult to identify. Soil is made up of horizons near the Earth's surface that have been transformed over time by climate, relief, and living organisms, as opposed to the underlying parent material. Soil commonly grades to hard rock or earthy materials at its lower limit, devoid of animals, roots, or other signs of biological activity. The bottom boundary of soil is arbitrarily defined at 200 cm for classification reasons (Pyle et al., 2018):

- The unconsolidated mineral or organic material that serves as a natural medium for the growth of terrestrial plants on the Earth's immediate surface.

- Unconsolidated mineral or organic matter on the Earth's surface that has been exposed to and displays the effects of genetic and environmental factors such as climate (including water and temperature effects), and macro- and microorganisms, conditioned by relief, acting on parent material over time. A product-soil differs from the material from which it is derived in many physical, chemical, biological, and morphological properties, and characteristics.

Solar

- Relating to or determined by the sun.
- Relating to or denoting energy derived from the sun's rays.

Solar Eclipse

A solar eclipse happens when the Moon casts a shadow over a section of the Earth, blocking sunlight completely or partially. When the Sun, Moon, and Earth are all aligned, this happens. A new moon (syzygy) occurs when such an alignment occurs, showing that the Moon is closest to the ecliptic plane. The disk of the Sun is completely hidden by the Moon during a total eclipse. Only a portion of the Sun is veiled in partial and annular eclipses (Press, 2008).

Total solar eclipses would occur every new moon if the Moon were in a perfectly circular orbit, a little closer to the Earth, and in the same orbital plane. The Moon's orbit, however, is inclined at more than 5° to the Earth's orbit around the Sun, thus its shadow frequently misses the Earth.

A solar eclipse can only happen during a new moon when the Moon is close enough to the ecliptic plane. Because the Moon's orbit crosses the ecliptic at its orbital nodes twice every draconic month (27.212220 days), while a new moon occurs once every synodic month, certain circumstances must occur for the two events to coincide (29.53059 days). Solar (and lunar) eclipses occur only during eclipse seasons, resulting in at least two and up to five solar eclipses every year, with no more than two total eclipses allowed (Plotnick et al., 2014).

Total eclipses are uncommon because the time of the new moon during the eclipse season must be precise in order for the viewer (on Earth) to align with the Sun and Moon's centers.

Solar Energy

Solar energy is solar radiation that can produce heat, cause chemical processes, or generate electricity. The overall amount of solar energy incident on Earth exceeds the world's current and projected energy needs by a large margin. This highly distributed source has the ability to meet all future energy needs if properly exploited. Solar energy, in contrast to the finite fossil fuels coal, petroleum, and natural gas, is predicted to become increasingly popular as a renewable energy source in the 21st century due to its limitless supply and nonpolluting nature (Pennington et al., 2020).

Although the Sun is a tremendously powerful energy source, and sunlight is by far the most abundant source of energy received by Earth, its intensity at the surface is quite low. This is primarily due to the vast radial spread of radiation from the faraway Sun (Figure 155).

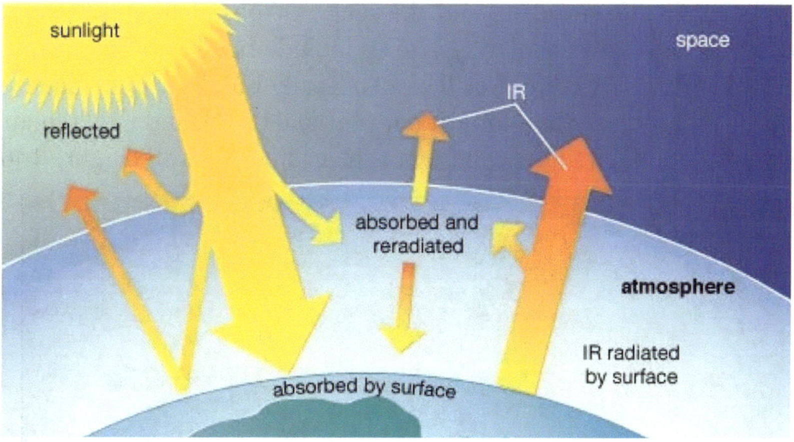

Figure 155. Solar energy.

Source: https://www.britannica.com/science/solar-energy.

The Earth's atmosphere and clouds absorb or scatter up to 54% of the incoming sunlight, resulting in a comparatively small additional loss. Nearly half of the sunlight that reaches the Earth is visible light, with the remaining 45% consisting of infrared radiation and smaller amounts of ultraviolet (UV) and other kinds of electromagnetic radiation.

Solar energy has huge potential, as Earth receives nearly 200,000 times the world's total daily electric-generating capacity in the form of solar energy every day. Unfortunately, despite the fact that solar energy is free, the high cost of collecting, converting, and storing it limits its use in many regions. Solar radiation can be turned into thermal energy (heat) or electrical energy, with the former being the more straightforward.

Solar System

The solar system is made up of the Sun, an average star in the Milky Way Galaxy, and the bodies that orbit it: 8 (formerly 9) planets with about 210 known planetary satellites (moons); countless asteroids, some with their own satellites; comets and other icy bodies; and vast reaches of highly tenuous gas and dust known as the interplanetary medium (Perdigão et al., 2020).

Ancient astronomers could see the Sun, Moon, and brightest planets with their naked eyes, and their observations and calculations of their movements gave origin to the discipline of astronomy. The amount of data on the motions, properties, and compositions of planets and smaller entities has exploded in recent years, and the range of observational sensors has expanded far beyond the solar system to include neighboring galaxies and the known universe's edge. Yet the solar system and its immediate outer boundary still represent the limit of our physical reach, and they remain the core of our theoretical understanding of the cosmos as well (Figure 156).

Figure 156. The eight planets of the solar system and Pluto, in a montage of images scaled to show the approximate sizes of the bodies relative to one another. Outward from the Sun, which is represented to scale by the yellow segment at the extreme left, are the four rocky terrestrial planets (Mercury, Venus, Earth, and Mars), the four hydrogen-rich giant planets (Jupiter, Saturn, Uranus, and Neptune), and icy, comparatively tiny Pluto.

Source: https://www.britannica.com/science/solar-system.

Solar Wind

Solar wind is a stream of particles, primarily protons and electrons, along with nuclei of heavier elements in lower numbers, that are propelled to velocities big enough to escape the Sun's gravitational field by the high temperatures of the solar corona. The solar wind is responsible for the Earth's magnetosphere's tail and comet tails, both of which point away from the Sun. During a relatively quiet period, at a distance of one astronomical unit (AU; the mean distance between Earth and the Sun, about 150 million km [93 million miles], the wind contains approximately 1 to 10 protons per cubic centimeter moving outward from the Sun at velocities of 350 to 700 km (about 220 to 440 miles) per second, resulting in a positive ion flux of 108 to 109 ions per square centimeter per second, each ion having an energy equal to at least 15 electron volts (Parkes and Morris, 2001). During solar flares, the proton velocity, flux, plasma temperature, and associated turbulence increase substantially.

Solid

A solid is one of the three basic forms of matter, the others being liquid and gas. (Sometimes, plasmas or ionized gases are considered a fourth state of matter.) Because the energy of atoms diminishes when the atoms take up a relatively ordered, three-dimensional structure, a solid form from a liquid or gas.

Solids differ from liquids and gases in several ways. For example, all solids can withstand forces applied perpendicular or parallel to a surface (i.e., normal or shear loads, respectively). Such qualities are determined by the properties of the atoms that comprise the solid, how those atoms are organized, and the forces that exist between them.

Solids are classified into three types: crystalline, noncrystalline (amorphous), and quasicrystal line. In a periodic atomic arrangement, crystalline solids exhibit a very high degree of order. This class includes almost all metals and numerous other minerals, such as common table salt (sodium chloride). Noncrystalline substances have atoms and molecules that are not arranged in a specific lattice pattern (Özyurt et al., 2021). Glasses, polymers, and gels are among them. Quasicrystalline solids exhibit novel symmetries in which the atoms are organized in quasiperiodic patterns that do not repeat at regular intervals. They have symmetries, such as fivefold symmetry, that are not found in conventional crystals. Quasicrystal formations are frequent in aluminum alloys containing another metal, such as iron, cobalt, or nickel.

Soluble

Solubility in chemistry refers to a material's capacity to create a solution with another substance, the solvent. The inability of the solute to produce such a solution is referred to as insolubility.

The extent of a substance's solubility in a certain solvent is commonly assessed as the concentration of the solute in a saturated solution, which can no longer dissolve any more solute. The two compounds are said to be in solubility equilibrium at this moment. There may be no such limit for some solutes and solvents, in which case the two compounds are said to be "miscible in any quantities" (or just "miscible").

The solute might be solid, liquid, or gas, whereas the solvent is often solid or liquid. Both can be pure substances or solutions in and of themselves.

Except in the most severe cases, gases are always miscible in all proportions, and a solid or liquid can only be "dissolved" in a gas by first entering the gaseous form.

Solute and solvent solubility are primarily determined by the makeup of the solute and solvent (including pH and the presence of additional dissolved substances), as well as temperature and pressure (Orion, 2019). The interaction between the particles (atoms, molecules, or ions) of the two substances, as well as thermodynamic concepts such as enthalpy and entropy, are frequently used to explain the dependency.

Solution

A solution is a form of homogenous mixture made of two or more components in chemistry. A solute in such a mixture is a substance that has been dissolved in another substance known as a solvent. The mixing of a solution occurs at a scale where the effects of chemical polarity are implicated, resulting in solvation-specific interactions. When the solvent is the larger proportion of the combination, as is often the case, the solution has the state of the solvent. The concentration of a solution is an important metric since it measures the amount of solute in a given amount of solution or solvent. When one of the solvents is water, the phrase "aqueous solution" is used (Figure 157).

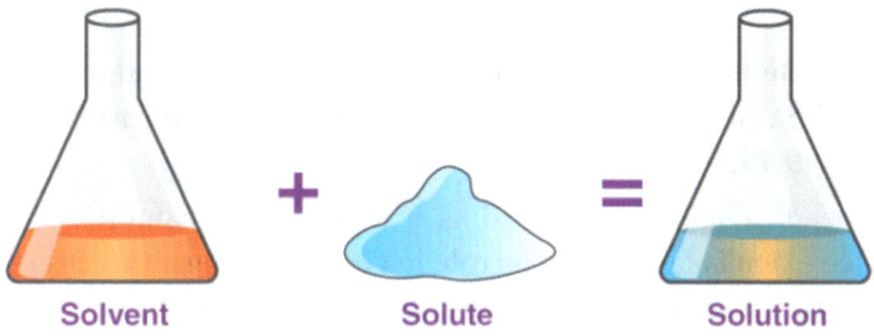

Figure 157. Solutions.

Source: https://byjus.com/chemistry/types-of-solutions/.

Sound Wave

Sound, like many other things, is a wave. Sound, in more technical terms, is a wave formed by air vibrations. When something generates a sound, it vibrates the air molecules, causing a chain reaction to occur that travels through the air until it reaches our eardrums. When our ears detect a sound, messages are delivered to our brain, which allows us to comprehend what we are hearing. So, according to that scientific definition, whether someone is present to hear it or not, the tree makes a sound!

When an object vibrates, it creates a pressure wave, which produces sound. This pressure wave generates vibrational motion in the particles in the surrounding media (air, water, or solid). When particles vibrate, they move surrounding particles, causing the sound to travel deeper across the medium. When vibrating air particles vibrate small components of the ear, the human ear detects sound waves (Oreskes, 2015).

Sound waves are similar to light waves in many respects. They both come from a certain source and can be distributed or dispersed in many ways. Sound waves, unlike light, can only travel through a medium such as air, glass, or metal. This implies that there is no sound in space!

Span

- The full extent of something from end to end; the amount of space that something covers.
- The length of time for which something lasts.

Speed of Light

The speed of light is the rate at which light waves travel through various materials. The value for the speed of light in a vacuum, in particular, is now defined as 299,792,458 meters per second.

The speed of light is regarded as a fundamental natural constant. Its significance extends far beyond its role in characterizing an electromagnetic wave characteristic. It acts as the universe's lone limiting velocity, acting as an upper limitation on signal transmission rates as well as the speeds of all material particles. In the classic relativity equation, $E = mc^2$, the speed of light (c) serves as a proportionality constant, connecting the previously distinct ideas of mass (m) and energy (e) (E) (Oreskes, 2013).

Sphere

(Geometry) A body or space contained within a single surface, each part of which is equally distant from a point within designated its center.

(Science: astronomy) The apparent surface of the heavens, assumed to be spherical and equally distant everywhere, on which the heavenly bodies appear to have their places, and on which the various astronomical circles, such as right ascension and declination, the equator, ecliptic, and so on, are conceived to be drawn; an ideal geometrical sphere, with the astronomical and geographical circles in their proper positions. One of the concentric and eccentric revolving spherical transparent shells in which the stars, sun, planets, and moon were assumed to be put and carried in such a way as to cause their apparent motions in ancient astronomy.

Split

- To split longitudinally; especially, to separate from end to end. Forcefully dividing in the direction of the grain layers; rive; cleave; as, splitting a piece of timber or a board; splitting a jewel; splitting a sheepskin. The rocks were split in two by the cold winter (Dryden).
- To burst, rupture, rend, or tear apart Congealed water shattered a massive vase of extremely hard marble (Boyle).
- To separate or divide into pieces or divisions, as by strife; to separate into sections or parties, as a political party; to disunite (Odenwald, 2020).

- (Science: chemistry) To divide or separate into components; frequently used with up; for example, to split sugar into alcohol and carbonic acid. To separate hairs, to make insignificant distinctions.

Spring Equinox

The spring or vernal equinox, which occurs twice a year when the Sun is directly above the Equator and day and night are of equal length; also, either of the two locations in the sky where the ecliptic (the Sun's annual course) and the celestial equator connect. The vernal equinox occurs in the Northern Hemisphere around March 20 or 21, when the Sun crosses the celestial equator heading north. The equinox happens in the Southern Hemisphere on September 22 or 23, when the Sun passes south across the celestial equator (Neeck and Volz, 2011). The vernal equinox also marks the start of spring, which lasts until the summer solstice, according to the astronomical definition of the seasons (June 20 or 21 in the Northern Hemisphere, December 21 or 22 in the Southern Hemisphere).

Squall Line

A squall line, also known as a quasi-linear convective system (QLCS), is a line of thunderstorms that develops along or ahead of a cold front. The term was used as a synonym for cold front in the early 20th century. It includes heavy rain, hail, frequent lightning, strong straight-line winds, and the possibility of tornadoes and waterspouts. Strong straight-line winds are possible where the squall line is shaped like a bow echo. Tornadoes can form along waves in a line echo wave pattern (LEWP) that contain mesoscale low-pressure zones. Some bow echoes that form throughout the summer are known as derechos, and they move quickly across wide areas of territory. A wake low, occasionally linked with a heat explosion, can be found on the back border of the rainband associated with mature squall lines (Figure 158) (Bohle, 2015).

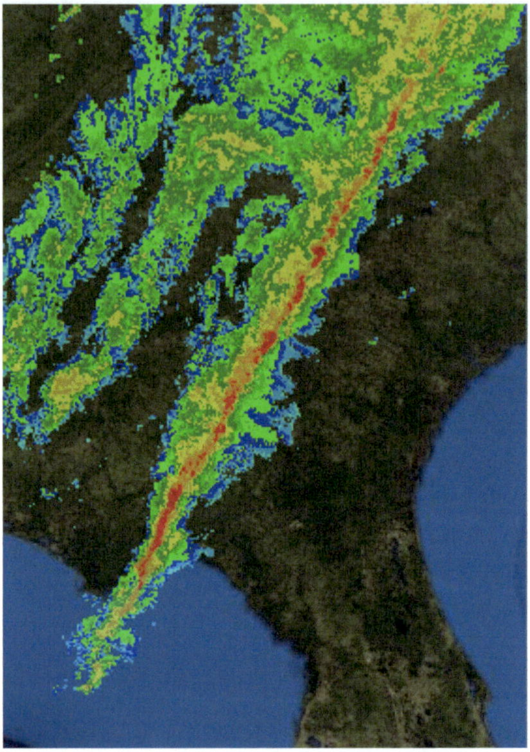

Figure 158. A squall line over 1,000 mi (1,600 km) long across the Gulf of Mexico and Eastern U.S. on 30 January 2013 (radar coverage is from ground radars, so middle image does not cover the portion over the Gulf). The rightmost image is a couple of hours after the other two, showing the strongest portion of the line as it passes through Florida, Georgia, and South Carolina.

Source: https://en.wikipedia.org/wiki/Squall_line#/media/File:Squall_line_South_Georgia_and_Florida_31_January_2016_0000UTC.png.

Stalactite

Stalactites are mineral formations that hang from the ceilings of caves, hot springs, or man-made structures like bridges and mines. A stalactite can be formed by any soluble material that can be deposited as a colloid, is in suspension, or can be melted. Lava, minerals, mud, peat, pitch, sand, sinter, and amberat can all be found in stalactites (crystallized urine of pack rats). Although speleothems are the most common type of stalactite due to the availability of limestone caverns, stalactites are not always speleothems (Movva et al., 2012).

A stalagmite is the name given to the equivalent formation on the cave floor. There are mnemonics for which word refers to which sort of formation; for example, stalactite has a C for "ceiling" and stalagmite has a G for "ground."

Stalagmite

A stalagmite is a type of rock structure that rises from the cave floor as a result of material deposited on the floor by ceiling drippings. Stalagmites are usually made of calcium carbonate, although they can also be made of lava, mud, peat, pitch, sand, sinter, and amberat (crystallized urine of pack rats).

A stalactite is the equivalent formation that hangs from the cave ceiling:

1. **Standard Pressure:** Standard atmospheric pressure, commonly known as barometric pressure, is the force exerted by an atmospheric column per unit area (that is, the entire body of air above the specified area). A mercury barometer (thus the often-used synonym barometric pressure) can be used to measure atmospheric pressure, which displays the height of a column of mercury that exactly balances the weight of the column of atmosphere over the barometer (Müller et al., 2019). An aneroid barometer, in which the sensing element is one or more hollow, partially evacuated, corrugated metal disks supported against collapse by an inside or outside spring, can also be used to measure atmospheric pressure; the change in shape of the disk with changing pressure can be recorded using a pen arm and a clock-driven revolving drum.

Atmospheric pressure is measured in a variety of quantities, including millimeters (or inches) of mercury, pounds per square inch (psi), dynes per square centimeter (dynes per square centimeter), millibars (mb), standard atmospheres, and kilopascals. By definition, standard sea-level pressure is 760 mm (29.92 inches) of mercury, 14.70 pounds per square inch, 1,013.25 103 dynes per square centimeter, 1,013.25 millibars, one standard atmosphere, or 101.325 kilopascals. The differences between these values are relatively minor; for example, the highest and lowest sea-level pressures ever recorded are 32.01 inches (in the center of Siberia) and 25.90 inches (in the Arctic Circle) (in a typhoon in the South Pacific). The Earth's wind and storm patterns are mostly determined by the minor differences in pressure that exist.

2. **Steep:** Having an almost vertical slope or pitch, or a relatively high gradient, as a hill, an ascent, stairs, etc.
3. **Stratified:** Stratification refers to the layering that occurs in most sedimentary rocks and igneous rocks generated at the Earth's surface, such as lava flows and volcanic fragmental deposits. The thickness of the strata ranges from a few millimeters to many meters, and their form varies substantially. Strata can range from thin sheets covering many square kilometers to thick lens like forms extending barely a few meters laterally (Metz, 2008).

Stratification planes are planes of parting or separation between individual rock layers. They are horizontal where sediments are deposited as level layers and incline where the depositional site was a sloping surface. The bottom surface of a stratum approximately conforms to the imperfections of the underlying surface; nevertheless, the stratification plane above the stratum is nearly level.

Stratification in sedimentary rocks can occur as a result of changes in texture or composition during deposition, or as a result of pauses in deposition that allow older deposits to alter before new sediments cover them. A stratigraphic sequence may thus appear as an alternation of coarse and fine particles, as a series of color changes caused by differences in mineral composition, or simply as layers of comparable aspect separated by various planes of separating (Li and Li, 2016). There is no direct relationship between stratum thickness and extent and deposition rate or time represented; for example, a stratum of limestone 2.5 cm (1 inch) thick may take longer to form than a stratum of sandstone 3 m (10 feet) thick. The most common source of stratification is difference in the depositing agent's carrying ability. Water and wind separate sediments based on particle size, weight, shape, and these sediments settle in layers of relative homogeneity.

Differences in sediment composition resulting from different sources, and variation in sediment brought about by change in agents of deposition, also lead to stratification.

Where strata have been distorted, the record of past motions of the Earth's surface is maintained in the stratification, allowing geologic events to be interpreted and practical consequences such as the location of mineral resources, petroleum fields, and groundwater reservoirs to be achieved.

Stratification in sedimentary rocks varies widely in terms of prominence and structural characteristics. In general, it develops best in fine-grained sediments and is least visible and permanent in coarse-grained materials like conglomerates (Ledley et al., 2011). Two unique structural kinds are recognized as being indicative of specific habitats. Cross-bedding, which is frequent in fluvial or eolian deposits, and graded bedding, which implies transport by density (or turbidity) currents or, in some circumstances, varved deposits, are two examples.

Stratification in volcanic rocks differs from that in sedimentary rocks in some ways. Gravity, particle size, and wind effect the separation of fragmented volcanic debris in flight. When it falls to the ground, it may form well-organized strata. If it falls into lakes or the sea, it layers up like any other watery detritus. Stratification can also occur as a result of successive liquid lava flows or alternations between flows and ashfalls.

Stratification does not exist in all sedimentary deposits. Those transported solely by ice, landslide deposits, and residual soils, for example, do not show any stratification. Plants and animals, recrystallization of limestones, and other post-depositional events can all damage original strata.

Stratosphere

A layer of the Earth's atmosphere known as the stratosphere. As you move upward, you will reach the second layer of the atmosphere. The troposphere, the lowest layer, is located directly underneath the stratosphere. The mesosphere is the next higher layer above the stratosphere.

At moderate latitudes, the floor of the stratosphere is around 10 km (6.2 miles or 33,000 feet) above the ground. The stratosphere's top occurs at an altitude of 50 km (31 miles). The height of the stratosphere's bottom varies with latitude and season. In winter, the stratosphere's lower border can be as high as 20 km (12 miles or 65,000 feet) near the equator and as low as 7 km (4 miles or 23,000 feet) in the poles. The tropopause is the lower limit of the stratosphere, and the stratopause is the upper boundary (Lewis, 2008).

Ozone, a rare form of oxygen molecule that is quite prevalent in the stratosphere, heats this layer by absorbing energy from the Sun's incoming UV light. Temperatures rise when one ascends into the stratosphere. This is the polar opposite of what happens in the troposphere, where temperatures drop as height increases. Because of the temperature stratification, there is little convection and mixing in the stratosphere, resulting in relatively stable air layers. Commercial jet aircraft operate in the lower stratosphere to avoid turbulence in the troposphere below.

The stratosphere is extremely dry, with very little water vapor in the air. As a result, few clouds form in this layer; practically all clouds form in the lower, more humid troposphere. The exception is polar stratospheric clouds (PSCs). In the winter, PSCs can be found in the lower stratosphere near the poles.

They are found at altitudes ranging from 15 to 25 kilometers (9.3 to 15.5 miles) and form only when temperatures fall below –78°C. They appear to contribute to the production of the famed ozone layer holes by "encouraging" specific chemical reactions that degrade ozone. PSCs are also known as nacreous clouds (Košler, 2007).

At the apex of the stratosphere, air is nearly 1,000 times thinner than at sea level. As a result, jet planes and weather balloons reach their maximum operational heights in the stratosphere.

Stratus

Stratus clouds are low-level clouds with horizontal layering and a homogenous base, as opposed to convective or cumuliform clouds created by rising thermals. More specifically, stratus refers to flat, hazy, featureless clouds at low altitudes that range in hue from dark gray to virtually white. The term stratus is derived from the Latin prefix strato-, which means "layer." Stratus clouds may cause a light shower or a dusting of snow. These clouds are essentially above-ground fog that emerge as a result of either the lifting of morning fog or cold air flowing at low altitudes over a region. Because of their fog-like appearance, these clouds are sometimes referred to as "high fog." While light rain may fall, this cloud does not predict considerable precipitation.

When a sheet of warm, moist air lifts from the ground and depressurizes at the lapse rate, stratus clouds arise. As a result of the adiabatic cooling, the relative humidity rises (Kleinhans et al., 2010).

Stratus clouds have the appearance of featureless gray to white sheets of cloud. Depending on the temperature, they can be made of water droplets, supercooled water droplets, or ice crystals.

Stratus nebulosus clouds are characterized by a featureless or nebulous veil or layer of stratus clouds that lacks distinguishing features or structure. They can be found at low elevations and are a good indicator of atmospheric stability, indicating continued stable weather. Stratus nebulosus can cause light rain, drizzle, or snowflakes. Stratus fractus clouds, on the other hand, have an uneven shape and formations that are plainly broken or ragged.

They are most commonly seen during the precipitation of significant rain-bearing clouds, such as nimbostratus clouds and cumulonimbus clouds, and are classed as pannus clouds. Stratus fractus can also occur alongside mountain slopes in the absence of nimbus clouds (precipitating clouds), and its color ranges from dark gray to virtually white.

Stream

A stream is a body of water that runs on the surface of the Earth. The terms stream and river are frequently used interchangeably, however rivers typically denote bigger streams.

Streams bring numerous advantages to people. Streams, in addition to providing drinking water and irrigation for crops, can also wash away garbage and generate electricity via hydropower (Klippel et al., 2019). Streams are frequently used for leisure purposes, such as swimming, fishing, and boating. Streams also serve as vital habitat for wildlife.

Streams require two things in order to exist: gravity and water. When precipitation falls on the ground, some of it trickles into groundwater, but the majority of it runs downhill as runoff and gathers in streams. A watershed, also known as a drainage basin, is the area where water is collected for a stream. Smaller streams frequently join to generate larger streams as they travel downward. These smaller streams are referred to as tributaries.

Streams form channels by eroding rock and transporting it and other debris downstream. This is referred to as erosion. Backyard gullies are formed by the same forces that carved out the Grand Canyon.

Streams vary in shape depending on the geography through which they run. One canal crawl over the landscape of meandering streams. These curves can get so wide over time that they meet and close off the bend from the rest of the stream, forming oxbow lakes. Braided rivers feature numerous channels that divide and merge in the same way that a braided hair braid does. Planetary scientists have discovered landforms on Mars' surface that resemble stream channels, implying that water flowed there in streams at some point in the planet's history.

Streams support a diverse ecology of plants and animals, including fish, birds, salamanders, snails, and mammals.

Streams have developed to support a variety of flora and animals. Strong root systems that act as anchors are common in plants, as are long, flexible stems that may flow with the current. Fly larvae feed on fallen leaves in the water, and the flies themselves become fish food later on. Mussel's burrow into the sediment on the stream's bottom to stay put while filtering the water for food. Many fish have streamlined designs that allow them to stay in place and avoid being carried away by powerful currents (Hundman and Mattmann, 2017).

Many things can wreak havoc on the health of a stream. Dams obstruct the natural flow of sediment and water, preventing wildlife, nutrients, and water from freely flowing through the stream system. Untreated sewage from cities is sometimes discharged into waterways, generating algae blooms that smother wildlife. Pollutants from farms and industry can wash into streams, harming wildlife and those who rely on the water.

Structure

The earth's structure is separated into four major components: the crust, mantle, outer core, and inner core. Each layer has its own chemical composition and physical state, and it can have an impact on life on Earth's surface. Plates shift as a result of movement in the mantle induced by fluctuations in heat from the core, which can result in earthquakes and volcanic eruptions. These natural disasters then alter our terrain and, in certain situations, endanger life and property. With the help of these educational tools, you may learn more about how the world is built.

The study of the three-dimensional distribution of rock units in relation to their deformational histories is known as structural geology. The basic purpose of structural geology is to use measurements of present-day rock geometries to learn about the history of deformation (strain) in the rocks and, ultimately, to comprehend the stress field that resulted in the observed strain and geometries (Guo et al., 2016).

This understanding of the dynamics of the stress field can be linked to significant geologic events; a common goal is to understand the structural evolution of a specific area in relation to regionally widespread patterns of rock deformation (e.g., mountain building, rifting) caused by plate tectonics (Figure 159). The internal structure of Earth, also known as the structure of the solid Earth, or simply the structure of Earth, refers to the concentric spherical layers that divide the Solid Earth, omitting the atmosphere and hydrosphere. It is made up of a silicate solid crust on the outside, a highly viscous asthenosphere and solid mantle on the inside, a liquid outer core whose flow generates the Earth's magnetic field, and a solid inner core (Butler, 2016).

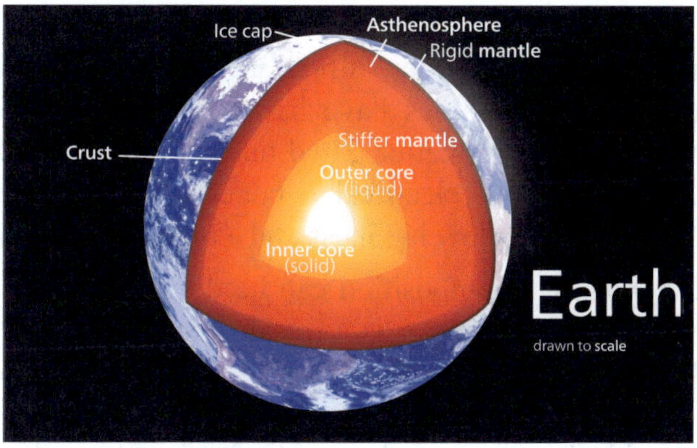

Figure 159. Structure of earth.

Source: https://en.wikipedia.org/wiki/Structure_of_Earth.

The scientific understanding of Earth's internal structure is based on observations of topography and bathymetry, observations of rock in outcrop, samples brought to the surface from greater depths by volcanoes or volcanic activity, analysis of seismic waves passing through Earth, measurements of Earth's gravitational and magnetic fields, and experiments with crystalline solids at pressures and temperatures typical of Earth's deep interior.

Subsurface

Earth material (such as rock) that is close to but not on the ground's surface. Subsurface structures are subsurface systems that collect runoff and slowly permeate it into groundwater via rock and gravel. A variety of subterranean infiltration systems can be constructed to improve groundwater recharge. Pre-cast concrete or plastic pits, chambers constructed pipes), perforated pipes, and galleys are the most popular varieties.

Geology and correlation of rock formations, structures, and other characteristics beneath the surface of the land or sea, as exposed or inferred by exploratory drilling, subterranean workings, and geophysical approaches.

The study of the physical qualities and location of rock and soil found beneath the ground surface is known as subsurface geology. Understanding the materials beneath man-made buildings is one of the most valuable reasons for learning about the subsurface. These studies are commonly called geotechnical reports and are critical for building structures safely in areas with geologic hazards.

Sulfide

Sulfide mineral, sulfide, also written sulfide, any member of a group of sulfur-metal compounds. Most sulfides have basic structural shapes, excellent symmetry in their crystal formations, and numerous metal characteristics, such as metallic luster and electrical conductivity. They are frequently brightly colored and have a low hardness and a high specific gravity.

The sulfide minerals' composition can be described using the general chemical formula AmSn, where A is a metal, S is sulfur, and m and n are integers, yielding A2S, AS, A3S4, and AS2 stoichiometries. Iron, copper, nickel, lead, cobalt, silver, and zinc are the most common metals found in sulfides, but roughly 15 other metals can also be found in sulfide structures (Goldstein et al., 2017).

Almost all sulfide minerals have structural configurations that fall into one of six categories, four of which are significant. These arrangements are close-packing combinations of metal and sulfur, governed by ionic size and charge.

The sodium chloride structure is the simplest and most symmetrical of the four major structural types, with each ion occupying a spot within an octahedron of six oppositely charged neighbors. Galena (PbS), a lead ore mineral, is the most common sulfide crystallizing in this manner. The pyrite structure is a kind of packing that contains two sulfide ions in each of the octahedral locations in the sodium chloride structure. This is a high-symmetry structure seen in the iron sulfide pyrite (FeS_2O). Sphalerite (ZnS) has a second unique structural type in which each metal ion is surrounded by six oppositely charged ions grouped tetrahedrally. The fluorite structural type is the third most important, in which the metal cation is surrounded by eight anions, each of which is surrounded by four metal cations.

The antifluorite structure is the inverse of this structure, with the metal cation surrounded by four anions and each anion surrounded by eight metal cations. It is a combination of some of the most valuable precious metal tellurides and selenides, including hessite (Ag_2Te), a silver ore mineral (Escorihuela, 2018).

The connection in almost all sulfides is covalent, but some have metallic characteristics. Sulfur's covalent property allows for the formation of sulfur-sulfur bonds and the inclusion of S_2 pairs in some sulfides, such as pyrite.

Layer structures are found in several sulfides, notably molybdenite (MoS_2) and covellite (CuS). The spinel (q.v.) structure is seen in several uncommon sulfide types.

Surface

In chemistry and physics, the surface is the outermost layer of a material or substance. Because the particles (atoms or molecules) on the surface have nearest neighbors alongside and below but not above, the physical and chemical properties of a surface differ from those of the bulk material; consequently, surface chemistry is a subfield of physical chemistry (Duffy, 2014). Aspects of surface behavior include crystal formation, the effects of catalysts and detergents, and the phenomena of adsorption, surface tension, and capillarity. Surface appearance is vital, whether produced through electroplating, paint, oxidation-reduction, bleaching, or another method.

T

Tectonic

The scientific study of the deformation of the rocks that make up the Earth's crust, as well as the factors that cause such deformation, is known as tectonics. It is concerned with the folding and faulting related with mountain formation, as well as the large-scale, progressive upward and downward movements of the crust (epeirogenic movements) and rapid horizontal displacements along faults. Igneous processes and metamorphism are two more phenomena explored (Clowes, 2010). Plate tectonics (q.v.) is the main working principle of tectonics, a theory developed in the late 1960s by American, Canadian, and British geophysicists to widen and synthesize the concept of continental drift and the seafloor spreading hypothesis.

Tectonics refers to the mechanisms that govern the structure and properties of the Earth's crust, as well as its evolution over time. These include the processes of mountain formation, the growth and behavior of the strong, old cores of continents known as cratons, and the interactions of the comparatively rigid plates that make up the Earth's outer shell. Tectonics also provides a framework for understanding earthquake and volcanic belts, which directly affect a large portion of the world's population.

Economic geologists hunting for fossil fuels and mineral deposits of metallic and nonmetallic resources rely on tectonic studies as guides. Geomorphologists must grasp tectonic concepts in order to explain erosion patterns and other Earth surface phenomena.

The outermost half of the Earth – the crust and uppermost mantle – is seen as a single mechanical layer, the lithosphere, in plate tectonics. The lithosphere is separated into different "plates" that shift relative to each other on the underlying, comparatively weak asthenosphere, a process that is ultimately driven by the constant loss of heat from the Earth's interior. Plate boundaries are classified into three types: divergent, where plates move apart and new lithosphere is created as a result of sea-floor spreading; transform, where plates slide past each other; and convergent, where plates converge and lithosphere is "consumed" by the process of subduction. Convergent and transform boundaries are the biggest structural discontinuities in the lithosphere and are responsible for the majority of major (Mw > 7) earthquakes around the planet. Convergent and divergent boundaries are also the site of most of the world's volcanoes, such as around the Pacific Ring of Fire. Most of the deformation in the lithosphere is related to the interaction between plates, either directly or indirectly (Figure 160).

Figure 160. Site of the boundary between the North American Plate and the Eurasian Plate, Iceland.

Source: https://www.britannica.com/science/tectonics.

Temperature

Temperature is the degree to which a thing is hot or cold. We are talking about temperature when we talk about anything feeling hot (like the soup we drink when we are ill) or cold (like the snow, especially if you are not wearing gloves) (Clary and Wandersee, 2011).

The temperature of a thing, which is commonly measured in degrees Fahrenheit or degrees Celsius, tells us how much heat, or energy, it has. A boiling cup of water has extremely active molecules that move very quickly, causing the heat we feel on our hands and faces. Colder items do not have the same amount of energy. Their molecules are far less active.

Consider this, when you put an oven mitt over a hot item that has just come off the stove, the oven mitt usually picks up some of the heat, right? This is one of the core concepts of thermodynamics, which analyzes (partly) how energy and heat interact (Figure 161).

Figure 161. A thermometer is a device used for measuring temperature. This ice-covered thermometer shows that the temperature is about 0°C, or 32°F. If this thermometer displayed the kelvin scale, used by scientists, the temperature would read about 274.

Source: https://www.nationalgeographic.org/encyclopedia/temperature/.

If no heat is exchanged between two objects, they are both the same temperature. However, if one object is hotter than another, heat will naturally transfer from the hotter object to the colder one; this is why the oven mitt becomes heated.

Water is commonly used by scientists to compare temperatures since it is simple to remember the boiling and freezing points of good-old H_2O: Water freezes at 0°C and boils at 100°C (Cossu et al., 2010). The Fahrenheit scale is a little more difficult to remember, but water boils at 212°F and freezes at 32°F.

Except for Belize, Myanmar, Liberia, and the United States, the Celsius scale is used in the majority of the world. The Kelvin scale, which does not measure temperature in degrees, is used by scientists. Zero Kelvin, commonly known as absolute zero, is the coldest temperature and has the lowest energy level. Absolute zero is equal to about –273°C.

Temporarily

For a limited period of time; not permanently.

Terminus

A terminal moraine, also known as an end moraine, is a type of moraine that occurs at the terminus (edge) of a glacier, indicating the glacier's maximum progress. Debris generated by plucking and abrasion that has been pushed by the front edge of the glacier is no longer propelled any further and is instead deposited in an unsorted mass of sediment. Because the glacier behaves similarly to a conveyor belt, the longer it remains in one location, the more material is deposited. The moraine is left as a reminder of the ice's final extent.

The surrounding land is constantly degrading as a glacier travel along its route. Loose rock and bedrock fragments are regularly scooped up and moved by the glacier. Fine sediments and particles are also mixed in with the glacial ice. When these rocks and sediments accumulate, they form what is known as glacial till (Cerase, 2019).

When a glacier retreats from a previously deposited terminal moraine, it pushes proglacial material or till into an existing terminal moraine. This technique has the potential to significantly increase the size of the present terminal moraine.

Dump moraines form when rock, silt, and debris that gather at the glacier's upper surface slip, fall, or flow from the glacier's snout. As the glacier recedes, the deposit of till will produce a terminal moraine.

Ablation moraines arise when a big chunk of ice, carrying silt and debris, breaks away from the glacial snout. The debris found throughout this glacier component is deposited to form a new terminal moraine once it is detached and begins to melt. It is vital to note that the more debris detected within the ice, the longer it will take for the ice to completely melt.

Climate has a significant impact on the creation of terminal moraines. As temperatures rise, glaciers begin to recede quicker, depositing more glacial till in the form of terminal moraines. When temperatures drop, though, the zone of accumulation goes into overdrive. This initiates a process in which the accumulation of snow in the accumulation zone exceeds the loss due to melting or ablation.

Terrain

The vertical and horizontal dimensions of the land surface are referred to as terrain or relief (see topographical relief). Bathymetry refers to underwater relief, whereas hypsometry investigates terrain in relation to sea level. Terra (the Latin origin of terrain) signifies "earth."

The lay of the land is referred to as topography in physical geography. This is typically described in terms of terrain features' elevation, slope, and direction. Surface water flow and distribution are influenced by terrain. It has the potential to influence weather and climate patterns over a wide area (Britner, 2008).

Understanding terrain is important for a variety of reasons, including:

- A region's viability for human habitation is largely determined by its terrain: flatter alluvial plains have better farming soils than steeper, rockier uplands.
- Comprehension the terrain of a place aids understanding of watershed boundaries, drainage characteristics, drainage systems, groundwater systems, water movement, and impacts on water quality in terms of environmental quality, agriculture, hydrology, and other interdisciplinary sciences. To estimate river water quality, complex arrays of relief data are utilized as input parameters for hydrological transport models (such as the SWMM or DSSAM Models).
- Understanding terrain also contributes to soil protection, particularly in agriculture. Contour plowing is a well-established practice that allows for sustainable agriculture on sloping land; it is the practice of plowing along lines of equal elevation rather than up and down a hill (Zhuang et al., 2019).
- Terrain is militarily important because it impacts armed forces' capacity to conquer and control regions, as well as move troops and material into and through them. Understanding terrain is fundamental to both defensive and offensive strategy. The military definition of "terrain" is quite broad, embracing not just landform but also land use and land cover, surface transportation infrastructure, physical structures, and human geography, and, by extension, psychological, cultural, or economic variables under the phrase human terrain.
- The terrain has a vital role in shaping weather patterns.

Because of height variations or a "rain shadow" effect, two sites geographically adjacent to each other may have drastically different precipitation levels or timing.

- Accurate topography information is essential in aviation, particularly for low-flying routes and maneuvers (see terrain collision avoidance) and airport heights. The range and performance of radars and terrestrial radio navigation systems will be influenced by terrain as well. Furthermore, a hilly or mountainous terrain can strongly impact the implementation of a new aerodrome and the orientation of its runways.

Theory

A theory is a collection of related ideas that are designed to explain something. A theory provides a framework for explaining observed phenomena. Assumptions underpin the explanations. A number of hypotheses can be derived from the assumptions. They can be put to the test to either support or refute the idea (Yu and Gong, 2012).

There are multiple interpretations for the term "theory":

- an estimate or a hunch;
- a law concerning entities that cannot be immediately observed, such as electrons or evolution;
- a comprehensive set of laws and hypotheses that explain a wide range of phenomena;
- a subject of study.

These interpretations can "shade into each other." Another meaning suggests:

- Hypothesis, but not confirmed;
- Knowledge that is systematically organize;
- Explanation.

A scientific theory (as opposed to a layman's theory) is "a logical, systematic collection of principles or explanation that has been verified—has stood up to attempts to show it untrue." For example, Darwin's theory of Evolution is a system of ideas that points to humans and apes having evolved from a common ancestor. This conclusion is based on evidence that supports it (Clowes, 2010).

Thermal

A thermal column (or thermal) is a rising air column at the lower altitudes of the Earth's atmosphere, a type of atmospheric updraft. Thermals are caused by the uneven heating of the Earth's surface caused by solar radiation, and they are a type of convection, specifically atmospheric convection. The Sun heats the ground, which warms the air directly above it. A thermal is a rising mass of buoyant air in the atmosphere, a convective circulation that transmits heat energy vertically (Witze, 2013).

Warmer air near the surface expands and becomes less dense than surrounding air. Because of its expansion in the lower pressure at higher altitudes, lighter air rises and cools. When it has cooled to the same temperature as the surrounding air, it stops rising.

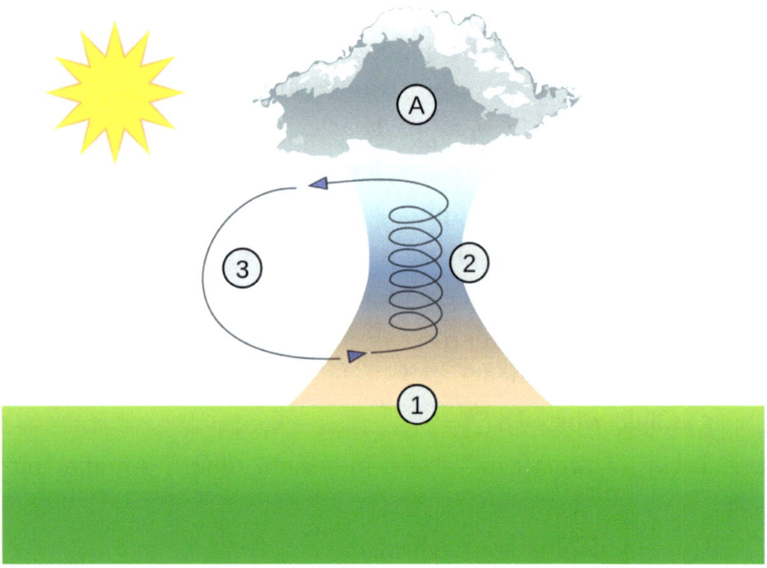

Figure 162. Example of a thermal column between the ground and a cumulus.

Source: https://en.wikipedia.org/wiki/Thermal.

A downward flow encircling a thermal column is associated with it. The outside is moving downward due to the displacement of cooler air at the top of the thermal.

Thermals can also be found in other parts of the solar system. Thermals are frequently visible in the form of dust devils on Mars, as seen from the rovers and orbiters.

Thermals can also be seen on the sun, where they often form hexagonal prisms (Bénard cells). In different situations, the phrase "thermal energy" is employed differently, and often vaguely. It relates to a number of distinct physical notions, including the internal energy, or enthalpy, of a body of matter and radiation; heat, defined as a sort of energy transfer (as is thermodynamic work); and the characteristic energy of a degree of freedom (Figure 162).

Thermometer

A thermometer is a device that measures temperature or the gradient of temperature (the degree of hotness or coldness of an object). A thermometer has two important components: (1) a temperature sensor (e.g., the bulb of a mercury-in-glass thermometer or the pyrometric sensor in an infrared thermometer) that changes as the temperature changes; and (2) a means of converting this change into a numerical value (e.g., the visible scale that is marked on a mercury-in-glass thermometer or the digital readout on an infrared model). Thermometers are frequently used to monitor processes in technology and industry, as well as in meteorology, medicine, and scientific study (Wilkinson et al., 2020).

Some thermometer principles were known to Greek philosophers 2,000 years ago. The thermometer's "development from a crude toy to an instrument of precision occupied more than a century, and its early history is encumbered with erroneous statements that have been repeated with such dogmatism that they have received the false stamp of authority," as Henry Carrington Bolton (1900) noted. The first thermometer was invented by the Italian physician Santorio-Santorio (Sanctorius, 1561–1636), but it was not standardized until the 17th and 18th century. Daniel Gabriel Fahrenheit achieved two important advancements in thermometry during the first decades of the 18th century in the Dutch Republic. He devised the mercury-in-glass thermometer (the first commonly used, accurate, and practical thermometer) as well as the Fahrenheit scale (first standardized temperature scale to be widely used).

Thunderstorm

Thunderstorms are powerful, short-lived meteorological disturbances characterized by lightning, thunder, dense clouds, heavy rain or hail, and strong gusty winds. Thunderstorms form when layers of warm, wet air rise to cooler regions of the atmosphere in a big, fast updraft. The liquid in the updraft condenses there, forming towering cumulonimbus clouds and, eventually, precipitation. Columns of cooled air then fall earthward, causing strong downdrafts and horizontal winds to strike the ground. Simultaneously, electrical charges accrue on cloud particles (water droplets and ice). When the stored electric charge becomes sufficiently big, lightning discharges occur (Wallace-Wells, 2019). Lightning heats the air it passes through so swiftly and fiercely that shock waves are formed; these shock waves are heard as thunderclaps and rolls. Severe thunderstorms are occasionally accompanied by spinning vortices of air that become concentrated and powerful enough to generate tornadoes.

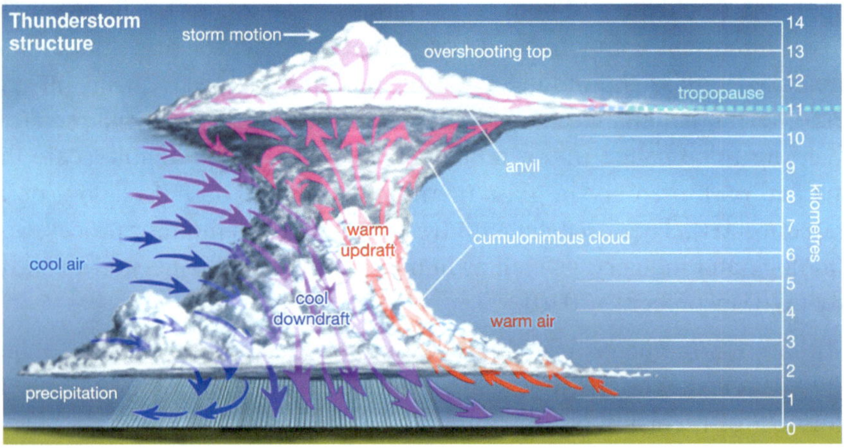

Figure 163. A towering thundercloud forms when the atmosphere gets unstable enough to create huge forceful updrafts and downdrafts (as illustrated by the red and blue arrows). Updrafts can be strong enough to push the cloud's top into the tropopause, the boundary between the troposphere (the lowest layer of the atmosphere) and the stratosphere. Click on the icons on the left side of the graphic to see representations of various thunderstorm-related events.

Source: https://www.britannica.com/science/thunderstorm.

Thunderstorms are known to occur in practically every region of the planet, while they are uncommon in the polar regions and uncommon at latitudes greater than 50° N and 50° S. As a result, thunderstorms are most common in temperate and tropical regions of the planet. The Florida peninsula (more than 80 thunderstorm days per year, with some locations exceeding 100), the Gulf Coast (60–90 days per year), and the New Mexico highlands (50–80 days per year) have the highest level of thunderstorm activity in the United States. Central Europe and Asia get 20 to 60 thunderstorm days per year on average. It is believed that there are around 1,800 thunderstorms in progress around the planet at any given time.

Large bands of rising and descending air are involved in the majority of brief but intense disruptions in Earth's wind systems. Thunderstorms are not immune to this tendency. Technically, a thunderstorm forms when the atmosphere becomes "unstable to vertical motion." When relatively warm, light air is overlain by cooler, heavier air, an instability like this can occur. Under these conditions, cooler air sinks, displacing warmer air upward. An updraft (a strong current of rising air) is formed when a sufficiently enough volume of air rises. If the updraft is moist, water condenses and forms clouds; condensation, in turn, releases latent heat energy, feeding upward air motion and creating instability (Figure 163).

Tidal

Tidal; caused by tides; having tides; rising and falling on a regular basis, or following and ebbing; as, tidal waters. "The tidal wave of deeper souls rolls into our inmost being, and takes us unawares from all meaner cares."

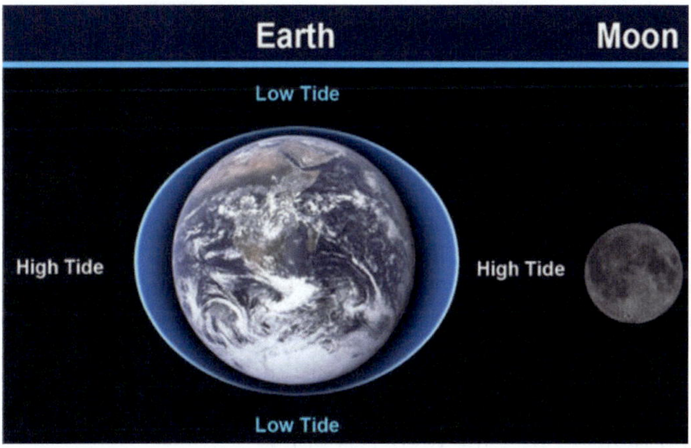

Figure 164. The gravitational attraction of the moon to ocean water creates the high and low tides.

Source: https://courses.lumenlearning.com/earthscience/chapter/tides/#:~:text=Tides%20are%20the%20daily%20rise,secondary%20cause%20(Figure%20below).

Tides are the daily rise and fall of sea level in a certain location. The primary cause of tides is the draw of the Moon's gravity on Earth, while the secondary reason is the pull of the Sun's gravity on Earth (Figure 164). The Moon has a larger impact because, despite being much smaller than the Sun, it is much closer. The Moon's gravitational force is around twice that of the Sun's.

To comprehend tides, it is best to begin with the Moon's effect on Earth. The Moon's gravity pushes Earth toward it as it orbits around our planet (Toth et al., 2014).

The lithosphere cannot move much, yet gravity pulls on the water, causing a bulge to form. The high tide beneath the Moon is represented by this bulge. The gravity of the Moon then pushes the Earth toward it, leaving the water on the other side of the planet behind. This causes a second high tide bulge on the Earth's opposite side from the Moon. The high tides are two water bulges on opposing sides of the Earth that are aligned with the Moon.

Tidewater

- Water overflowing land at flood tide also, water affected by the ebb and flow of the tide;
- Low-lying coastal land;
- Tidewater glaciers are valley glaciers that flow all the way down to the ocean. They often calve numerous small icebergs, which can sometimes pose problems for shipping lanes.

Tilt

The inclination angle of a planet's rotational axis in reference to a perpendicular to its orbital plane is referred to as axial tilt in astronomy.

It is also referred to as axial tilt or obliquity. The axial tilt is defined as the angle formed between the planet's axis and a line drawn perpendicular to the orbital plane across the planet's center.

The planet's axial tilt can also be described in terms of its orbital plane and a plane perpendicular to its axis. Because the Earth's orbital plane is known as the ecliptic in our solar system, the axial tilt of the Earth is formally known as the obliquity of the ecliptic.

The Earth's axial tilt is around 23° 27.' The axis is tilted in the same direction throughout a year; however, as the Earth orbits the Sun, the hemisphere (half part of earth) tilted away from the Sun will gradually come to be tilted towards the Sun, and vice versa (Tollefson, 2015).

This effect is the main cause of the seasons (see effect of Sun angle on climate). Whichever hemisphere is currently tilted toward the Sun experiences more hours of sunlight each day, and the sunlight at midday also strikes the ground at an angle nearer the vertical and thus delivers more heat.

Topography

The study of the forms and features of land surfaces is known as topography. The topography of a region may relate to the surface forms and features itself, or it may refer to a description of those forms and characteristics (especially their depiction in maps). Topography is a branch of geology and planetary science concerned with local detail in general, covering not only relief but also natural and man-made characteristics, as well as local history and culture (Thessen, 2016). In the United States, where topographic maps with elevation contours have rendered topography synonymous with relief, this definition is less prevalent (Telling et al., 2017). In a nutshell, topography is the recording of relief or terrain, the three-dimensional quality of the surface, and the identification of distinct landforms; this is also referred to as geomorphometry. In modern usage, this entails the creation of digital elevation data (DEM). It is often considered to include the graphic representation of the landform on a map by a variety of cartographic relief depiction techniques, including contour lines, hypsometric tints, and relief shading.

Topographic maps depict the flow of rivers, the rising of towering mountains, and the descent of steep valleys. They lay out the land, as shown in the following examples.

Topographic maps are used by engineers to plan a road, build a cell tower, or design a hydroelectric dam:

- Topography is used by geologists to understand tectonic activity, landforms, and where to dig a mine;
- Hikers use topographic maps to locate paths and the steepness of a slope in order to plan their ascent;
- Astronomers examine the geography of bodies other than Earth, such as the moon, Mars, or an asteroid;
- Climate scientists use topography to recognize air and water flow in climate models.

Topographers confront an uphill battle for accuracy and completeness as landscapes change and technology develops.

Topsoil

Topsoil is the soil's upper, outermost layer, often the top 5–10 inches (13–25 cm). It has the highest concentration of organic materials and microorganisms and is the site of the majority of the Earth's biological soil activity. Mineral particles, organic materials, water, and air make up topsoil. The amount of organic matter varies depending on the soil. With the presence of organic matter, the strength of the soil structure reduces, resulting in weak bearing capacity. Under some situations, such as roadbeds and foundations, organic matter condenses and settles in various ways. When the soil becomes dry, the structure suffers (Showstack, 2008). The volume of dehydrated topsoil falls significantly and may be eroded by wind. There is generally a high concentration of roots in this layer as plants obtain much of their vital nutrients here. Accurate depth of the topsoil layer is measured as the depth of the surface to the first densely packed soil layer known as subsoil.

Topsoil creation is a time-consuming process that normally takes 100 years for every inch of soil. It is generated as a result of rock weathering and the subsequent addition of organic material from rotting plants and animals. This improves the soil and adds nutrients necessary for plant life.

Topsoil creation is a time-consuming process that normally takes 100 years for every inch of soil. Soil composition and structure vary greatly from place to place, and it is constantly influenced by its surroundings. People discuss having clay or sandy soils in their gardens. Clay soils are more difficult to work with because they harden and the nutrients become trapped in the sticky soil, but they hold a lot more water (Bohle, 2015). However, sandy loam soil is the best and most commonly used in gardening projects. This soil has a deep, rich color, lightweight, and friable. As a quick test it should be hard to roll up into a ball, and then quick to break up.

Tract

Land lot or tract, a section of land.

Trigger

An event that causes a landslide event. Water is a common trigger.

Tropical

The tropics are the area of Earth that surrounds the Equator. They are separated by the Tropic of Cancer in the Northern Hemisphere (23°26′11.2″ (or 23.43644°) N and the Tropic of Capricorn in the Southern Hemisphere (23°26′11.2″ (or 23.43644°) S; these latitudes correlate to the Earth's axial tilt (Schmitz, 2011). The tropics are also known as the tropical zone or the torrid zone (see geographical zone). The tropics include any location on Earth that is a subsolar point (the Sun is directly overhead) at least once a year. As a result, the greatest latitudes of the tropics have the same positive and negative value. Similarly, because the Earth is not a perfect sphere, they approximate the "angle" of the Earth's axial tilt.

The "angle" is not entirely set, owing primarily to the effect of the moon, but the tropics are a geographic convention, being an averaged form, and the deviation is quite tiny.

In terms of climate, the tropics receive more direct sunshine than the rest of the planet and are generally hotter and wetter. The term "tropical" is occasionally used to refer to this type of climate rather than the geographical zone. Deserts and snow-capped mountains are included in the tropical zone, but they are not tropical in the climatic sense. The tropics are distinguishable from the other climatic and biophysical zones of the Earth, which are the middle latitudes and the polar regions on either side of the equator.

The tropics cover 40% of the Earth's surface area and encompass 36% of the Earth's landmass. As of 2014, the region was home to 40% of the world's population, and this figure was then projected to reach 50% by 2050 (Schmitz, 2012).

The tropics are warm all year, with temperatures averaging 25 to 28°C (77 to 82°F). This is because the tropics receive more Sun exposure. Because of that sun, the tropics do not have the same seasons as the rest of the world. The tropical seasons are divided into two categories: wet season and dry season.

The amount of rain that falls in the tropics varies dramatically from one region to the next. Some places, such as parts of South America's Amazon Basin, receive about 3 meters (9 feet) of rain every year. Other parts of the tropics have drier weather. The Sahara Desert in northern Africa receives only 2–10 millimeters (.793.9 inches) of rainfall per year.

The amount of rain that a location receives in the tropics has a direct impact on which plant and animal species reside there. The baobab tree thrives in the arid tropics of Africa, for instance. The baobab stores water in its huge trunk. On the other extreme is the rainy island of Sri Lanka in the

Indian Ocean. Sri Lanka gets enough precipitation to support 250 species of frogs.

Troposphere

The troposphere is the Earth's lowest layer of atmospheric. The troposphere contains the majority of the atmosphere's mass (about 75–80%). The troposphere is home to the majority of clouds, and it is also where the majority of weather occurs. The troposphere is by far the wettest layer of the atmosphere, with relatively little moisture in the layers above (Figure 165).

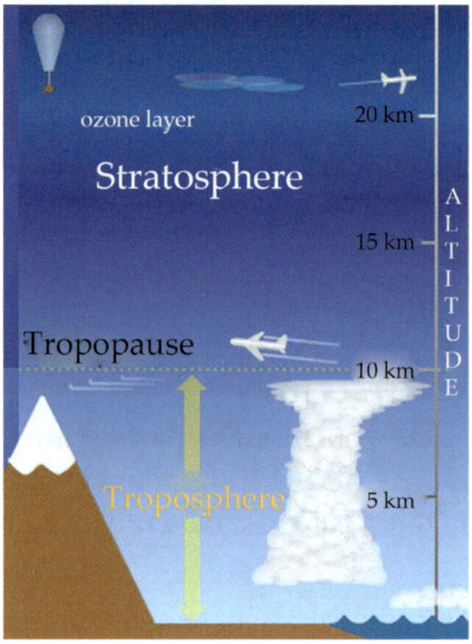

Figure 165. The troposphere.

Source: https://scied.ucar.edu/learning-zone/atmosphere/troposphere.

The troposphere's bottom is at the Earth's surface. The troposphere rises to a height of around 10 kilometers (6.2 miles or 33,000 feet) above sea level (Rao, 2012). The height of the troposphere varies with latitude (lowest in the poles, highest at the equator) and season (it is lower in winter and higher in summer). It can reach 20 km (12 miles or 65,000 feet) near the equator and 7 km (4 miles or 23,000 feet) over the poles in winter.

The air is hottest near the ground at the bottom of the troposphere. As one ascends into the troposphere, the air becomes cooler. As a result, even in the summer, the peaks of large mountains can be snow-covered.

Air pressure and density both drop when one ascends in altitude. That is why high-flying jet aircraft cabins are pressurized.

The stratosphere is the layer directly above the troposphere. The "tropopause" is the barrier between the troposphere and the stratosphere.

A liter of dry air in the Earth's planetary atmosphere contains 78.08% nitrogen, 20.95% oxygen, 0.93% argon, 0.04% carbon dioxide (CO_2), trace gases, and various amounts of water vapor. The sources of atmospheric water vapor are bodies of water (oceans, seas, lakes, rivers) on the planetary surface, which humidify the troposphere through evaporation, influencing the occurrence of weather phenomena; the greatest proportion of water vapor is in the atmosphere closest to the Earth's surface (Rivet and Kastens, 2012). The troposphere's temperature lowers at high altitude due to inversion layers that form at the tropopause, the atmospheric boundary that separates the troposphere from the stratosphere. The low air temperature at higher altitudes reduces the saturation vapor pressure, or the amount of atmospheric water vapor in the upper troposphere.

Tuff

Tuff is a porous, moderately soft rock created by the compaction and cementation of volcanic ash or particles. (In other cases, the Italian term tufa refers to a soft, porous sedimentary rock created by the chemical deposition of calcite, calcium carbonate, or silica from water as sinter.) Tuffs are classified as vitric, crystal, or lithic depending on whether they are mostly formed of glass, crystal chips, or the detritus of pre-existing rocks. The world's biggest quantities of vitric tuff are formed by eruptions through a large number of small fissures rather than volcanic cones.

Tuff can vary widely in texture, as well as chemical and mineralogical content, in large deposits. There has most likely been no geological period without volcanic eruptions; thus, tuffs range in age from Precambrian to Recent. Most of the older ones have lost all of their original textures and have been extensively recrystallized; many old basaltic tuffs are represented by green chlorite and hornblende schists, while many rhyolitic tuffs are represented by sericite schists (Figure 166) (Pyle et al., 2018).

In some eruptions, frothy magma wells to the surface as an emulsion of hot gases and incandescent particles; the shredded pumaceous material spreads quickly, even over gentle gradients, as a blazing avalanche (nuée ardente) that can move many kilometers at velocities exceeding 160 km/h.

After collapsing, the ejecta (erupted matter) may be securely compacted by adhesion of the hot glass particles to form streaky, welded tuffs (ignimbrites), such as those seen in New Zealand, Guatemala, Peru, and

Yellowstone National Park in the United States. When subsurface explosions occur, the fragmented material may be forcibly driven into the surrounding rocks, generating intrusive tuffs (peperites). Some of the world's largest deposits of vitric tuff are produced by eruptions through a large number of narrow fissures rather than from volcanic cones (Press, 2008).

Figure 166. Welded tuff from Bandelier National Monument, New Mexico.

Source: https://en.wikipedia.org/wiki/Tuff.

Turbulent

A turbulent flow is a type of fluid (gas or liquid) flow in which the fluid undergoes irregular fluctuations, or mixing, in contrast to laminar flow, in which the fluid moves in smooth paths or layers. Wind and water spinning around objects are common examples, as is any type of rapid flow (Reynolds number more than 2,100). Turbulence has properties such as eddies, vortices, and a reduction in drag. Golf balls with lower drag can travel farther than they would otherwise, and the dimpled surface of golf balls is designed to enhance turbulence in the boundary layer. If, as claimed, swimsuits with rough surfaces assist swimmers move quicker, the same explanation may apply.

In contrast to laminar flow, which moves in clean channels or layers, turbulent flow is a type of fluid (gas or liquid) flow in which the fluid experiences random fluctuations or mixing. In turbulent flow, the speed of the fluid at a given spot is constantly changing in magnitude and direction (Plotnick et al., 2014). Even though the currents are mild, the flow of wind and rivers is often tumultuous in this sense. The air or water swirls and eddies as the bulk of it goes in a certain direction.

Except for laminar flow near the leading edge of solids moving relative to fluids or extremely close to solid surfaces, such as the interior wall of a pipe, or fluids of high viscosity (relatively considerable sluggishness) flowing slowly through small channels, most fluid flows are turbulent. Common examples of turbulent flow are blood flow in arteries, oil transport in pipelines, lava flow, atmosphere, and ocean currents, the flow through pumps and turbines, and the flow in boat wakes and around aircraft-wing tips.

U

Ultraviolet (UV)

Ultraviolet (UV) radiation has wavelengths ranging from 10 nm (with a corresponding frequency of roughly 30 PHz) to 400 nm (750 THz), which is shorter than visible light but longer than X-rays. UV radiation is present in sunlight and accounts for around 10% of the total electromagnetic energy emitted by the Sun. Electric arcs and specialty lights, such as mercury-vapor lamps, tanning lamps, and black lights, can also produce it (Pennington et al., 2020). Although long-wavelength UV does not qualify as ionizing radiation because its photons lack the energy to ionize atoms, it can initiate chemical processes and cause many things to light or fluoresce. As a result, the chemical and biological effects of UV radiation outweigh the simple heating effects, and many practical applications of UV radiation stem from its interactions with organic molecules.

Short-wave UV radiation damages DNA while sterilizing surfaces with which it comes into contact (Figure 167).

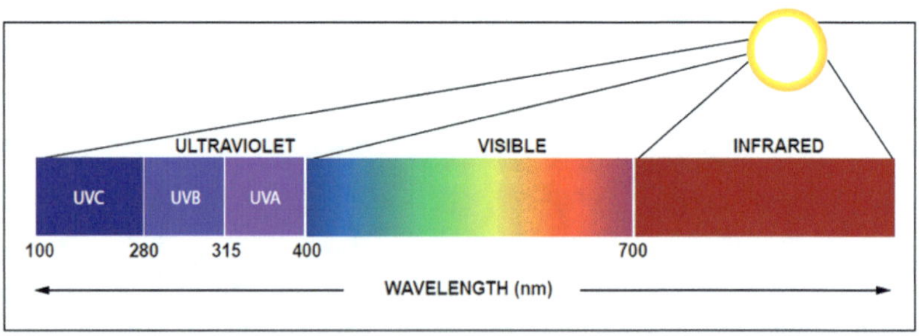

Figure 167. Ultraviolet radiation is made up of three types of rays – ultraviolet A, ultraviolet B, and ultraviolet C. Although ultraviolet C is the most dangerous type of ultraviolet light in terms of its potential to harm life on Earth, it cannot penetrate the Earth's protective ozone layer. Therefore, it poses no threat to human, animal, or plant life on Earth.

Source: https://eyescreen.cordlifetech.com/ultraviolet-radiation-harmful-or-harmless.

Suntan and sunburn are common side effects of skin exposure to UV rays, as also an elevated risk of skin cancer. Because of the amount of UV light produced by the Sun, the Earth would be unable to maintain life on dry land if the atmosphere did not filter away the majority of that light. Below 121 nm, more energetic, shorter-wavelength "extreme" UV ionizes air so intensely that it is absorbed before it reaches the Earth. UVB light, on the other hand, is responsible for the production of vitamin D in most land vertebrates, including humans. As a result, the UV spectrum has both positive and destructive impacts on life (Perdigão et al., 2020).

Because the lower wavelength limit of human vision is commonly assumed to be 400 nm, UV photons are invisible to humans, even though people can sometimes sense light at shorter wavelengths. Near-UV light can be seen by insects, birds, and some mammals.

Uneven

Not level or flat; spasmodic or variable; not parallel, straight, or horizontal.

Universe

The universe encompasses all of space and time, as well as their contents, which include planets, stars, galaxies, and all other forms of matter and energy. The Big Bang Theory is the most widely accepted cosmological explanation for the evolution of the universe. According to this idea, space, and time coexisted 13.7870.020 billion years ago and have been increasing ever since (Parkes and Morris, 2001). While the spatial size of the entire universe is unknown, the cosmic inflation equation predicts that it must have a minimum diameter of 23 trillion light-years, and the size of the observable universe, which is roughly 93 billion light-years in diameter at the present time, can be measured.

The earliest cosmological conceptions of the cosmos were geocentric, with Earth at the center, and were devised by ancient Greek and Indian philosophers. More precise astronomical measurements over the ages prompted Nicolaus Copernicus to establish the heliocentric hypothesis, which places the Sun at the center of the Solar System. Isaac Newton built on Copernicus' work, as well as Johannes Kepler's equations of planetary motion and Tycho Brahe's observations, in formulating the law of universal gravitation. Further advances in observational technology revealed that the Sun is one among hundreds of billions of stars in the Milky Way, which is one of a few 100 billion galaxies in the cosmos. Planets can be found around many of the stars in a galaxy. At the biggest scale, galaxies are dispersed uniformly and, in all directions, implying that the universe lacks both an edge and a core. At lower scales, galaxies are arranged in clusters and superclusters, which produce huge filaments and voids in space, resulting in a gigantic foam-like structure. Discoveries in the early 20th century revealed that the cosmos had a beginning and that space has been expanding at an accelerating rate since then (Özyurt et al., 2021).

According to the Big Bang theory, the original density of energy and matter has decreased as the universe has expanded. The cosmos gradually cooled and continued to expand after an initial accelerated expansion known as the inflationary period at roughly 1,032 seconds, and the separation of the four known fundamental forces, permitting the formation of the first subatomic particles and elementary atoms.

Under the influence of gravity, dark matter gradually accumulated, generating a foam-like structure of filaments and gaps. Giant clouds of hydrogen and helium were progressively pulled to the densest areas of dark matter, generating the first galaxies, stars, and everything else observed today.

The movement of galaxies has revealed that the universe contains far more matter than is accounted for by visible objects such as stars, galaxies, nebulae, and interstellar gas. This unseen stuff is referred to as dark matter (dark means that there is a wide range of strong indirect evidence that it exists, but we have not yet detected it directly).

The CDM model is the most frequently accepted universe model. It implies that approximately 69.2% 1.2% of the mass and energy in the universe is a cosmological constant responsible for the current expansion of space, and approximately 25.8% 1.1% is dark matter (Orion, 2019). As a result, ordinary ('baryonic') matter constitutes just 4.84% 0.1% of the physical world. Stars, planets, and visible gas clouds account for just around 6% of ordinary stuff.

There are numerous conflicting views concerning the eventual fate of the universe and what, if anything, preceded the Big Bang, while other physicists and philosophers refuse to speculate, believing that prior states will never be accessible. Some scientists have proposed numerous multiverse ideas in which our universe could be one of many other universes that also exist.

V

Valley

A valley is a long, narrow depression in the Earth's surface. Valleys are typically drained by rivers and can be found on a relatively level plain or between ranges of hills or mountains. Rift valleys are valleys formed by tectonic activity. Gorges are very narrow, deep valleys with a similar aspect. Both of these later forms are most commonly cut in flat-lying strata, but they can occur in other geological conditions as well.

Wherever there is enough rainfall, the land surface has the potential to grow into the known patterns of hills and valleys. There are, of fact, hyperarid areas with little fluvial activity. There are other geomorphological conditions in which the permeability of rocks or sediments causes so much infiltration that water cannot concentrate on the ground surface. Furthermore, some landscapes may be so new that there has not been enough time for fluvial action to modify them. The role of fluvial action on landscape, including long-term evolutionary processes, is considered here in detail. For additional information on fluvial and hillslope processes relating to valley formation (Oreskes, 2015).

The Kli Gandaki River valley in Nepal is the world's deepest subaerial valley. The valley is six kilometers long and is between two 8,000-meter (26,000-foot) Himalayan peaks, Dhaulagiri and Annapurna (four miles). Because the Himalayas are one of the most active tectonic uplift locations on the planet, this valley exemplifies the notion that the most rapid downcutting happens in areas with the most rapid uplift. The reason for this seeming paradox is found in the energetics of the degradation processes that characterize valley creation. As will be detailed further below, the steeper the gradient or slope of a stream, the more power it expands on the streambed (Butler, 2016). As uplift causes higher relief and steeper slopes, rivers have more erosional strength. As a result, the quickest relief reduction procedures might occur in areas of the most rapid relief generation (Figure 168).

Figure 168. The Colorado River flowing through the Grand Canyon.

Source: https://en.wikipedia.org/wiki/Grand_Canyon.

The Grand Canyon of the Colorado River in northern Arizona is perhaps the most famous example of a canyon. The Grand Canyon is around 1.6 km (1 mile) deep and 180 meters (590 feet) to 30 km (19 miles) wide, and it runs for 443 km (275 miles) over a vast upwarp of sedimentary strata carved by the Colorado River (Oreskes, 2013).

The incising motion of rivers creates the relief of valleys and canyons. Although hillslope processes are important in the formation of valley sides, rivers reduce erosion through deterioration. Rivers eventually settle to a base level, which is defined as the lowest point at which potential energy may be converted to kinetic energy of river flow. In most circumstances, sea level is the final base level for rivers.

Vapor

A vapor (American English) or vapor is a substance in the gas phase at a temperature lower than its critical temperature, which means it may be condensed to a liquid by raising the pressure on it without lowering the temperature. A vapor is not the same as an aerosol. An aerosol is a suspension of microscopic liquid, solid, or both particles in a gas.

The critical temperature of water, for example, is 647 K (374°C; 705°F), which is the greatest temperature at which liquid water can exist. At typical temperatures, gaseous water (known as water vapor) will condense into a liquid if its partial pressure is sufficiently elevated in the atmosphere (Britner, 2008).

A vapor and a liquid can coexist (or a solid). When this occurs, the two phases are in equilibrium, and the gas-partial pressure equals the liquid's equilibrium vapor pressure.

A vapor is a gas phase at a temperature below the critical temperature of a substance that can also exist in a liquid or solid-state. (For example, water has a critical temperature of 374°C (647°K), which is the greatest temperature at which liquid water can exist.) If the vapor comes into touch with a liquid or solid phase, the two will be in equilibrium. A compressible fluid phase is referred to as a gas. Fixed gases are gases in which no liquid or solid can form at the gas's temperature, such as air at normal ambient temperatures. A liquid or solid does not have to boil in order to produce vapor.

Vapor is in charge of the well-known processes of cloud formation and condensation. Prior to gas chromatography, it is widely used to perform the physical operations of distillation and headspace extraction from a liquid sample.

A vapor's constituent molecules move in three directions: vibrationally, rotationally, and translationally. The kinetic theory of gases takes these motions into account.

Vapor Pressure

At a given temperature, the vapor pressure is the equilibrium pressure of a liquid or solid. The amount of contact with the liquid or solid interface has no effect on the equilibrium vapor pressure of a liquid or solid.

A liquid's normal boiling point is the temperature at which the vapor pressure equals normal air pressure (Figure 169).

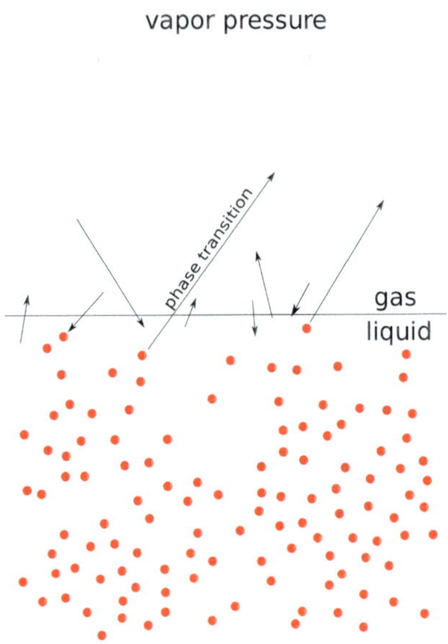

Figure 169. The microscopic process of evaporation and condensation at the liquid surface.

Source: https://en.wikipedia.org/wiki/Vapor_pressure.

The vapor pressures of the different phases are identical in two-phase systems (e.g., two liquid phases). In the absence of stronger inter-species affinities between like-like or like-unlike molecules, the vapor pressure obeys Raoult's law, which states that the partial pressure of each component is the product of its pure component's vapor pressure and its mole fraction in the mixture. The sum of the component partial pressures yields the overall vapor pressure.

Variation

In biology, variation refers to any difference between cells, individual organisms, or groups of organisms of any species that is generated by genetic differences (genotypic variation) or by the effect of environmental conditions on the manifestation of genetic potentials (phenotypic variation). Physical appearance, metabolism, fertility, mode of reproduction, behavior, learning, and mental aptitude, and other evident or measurable characteristics may differ (Odenwald, 2020).

Genotypic variances result from differences in the number or structure of chromosomes, as well as differences in the genes carried by the chromosomes. Genotypic variations include eye color, body shape, and illness resistance.

Polyploids are people who have several sets of chromosomes; many common plants have two or more times the normal number of chromosomes, and this type of variation can lead to the emergence of new species. A variation cannot be identified as genotypic by observation of the organism; breeding experiments must be performed under controlled environmental conditions to determine whether or not the alteration is inheritable.

Variations generated by the environment can be the consequence of a single component or the combined impacts of numerous factors, such as climate, food supply, and the actions of other creatures. Stages in an organism's life cycle and seasonal fluctuations in an individual are also examples of phenotypic variations. These variants do not include any hereditary changes and, in general, are not passed down to future generations; thus, they are insignificant in the evolutionary process.

Variations are classified as either continuous or quantitative (smoothly grading between two extremes, with the majority of individuals at the center, as height varies in human populations); or discontinuous or qualitative (smoothly grading between two extremes, with the majority of individuals at the center, as height varies in human populations) (composed of well-defined classes, as blood groups vary in humans). A polymorphic variation is a discontinuous variation with numerous classes, none of which are very tiny (Neeck and Volz, 2011).

Polymorphic variation can be seen in the separation of higher organisms into men and females, as well as the existence of numerous versions of the same species of butterfly, each colored to blend with a different vegetation. A genotypic variation cannot be identified by observing the organism; breeding studies under controlled environmental conditions are required to ascertain whether or not the alteration is inheritable.

Vary

The verb vary means to differ, change, diversify, or stray from the norm. Similarly, to vary means to change (something) such that it is not always the same. Very is an emphatic word that implies truly, definitely, or extremely. It is both an adjective and an adverb.

To change, alter, or modify one's appearance, character, form, attribute, etc., to be different or cause one's appearance, character, form, attribute, etc.

Velocity

The velocity of an item is a function of time and is defined as the rate of change of its position with respect to a frame of reference. Velocity is a specification of an object's speed and direction of travel (for example, 60 km/h to the north). The idea of velocity is crucial in kinematics, the part of classical mechanics that explains the motion of bodies.

Velocity is a physical vector quantity that requires both magnitude and direction to define. Speed is the scalar absolute value (magnitude) of velocity, and it is a coherently derived unit whose quantity is measured in the SI (metric system) as meters per second (m/s or ms^{1}). "5 meters per second," for example, is a scalar, whereas "5 meters per second east" is a vector. When there is a change in speed, direction, or both, the item is said to be accelerating.

Vent

In Earth science, a vent is a fissure on the seafloor through which geothermally heated water flows. Hydrothermal vents are typically found near volcanically active locations, spreading centers, ocean basins, and hotspots, as well as sites where tectonic plates are moving apart. The operation of hydrothermal vents creates rocks and mineral ore deposits known as hydrothermal deposits (Movva et al., 2012).

Because the Earth is both geologically active and possesses enormous volumes of water on its surface and within its crust, hydrothermal vents exist. Hydrothermal vents may generate features known as black smokers or white smokers beneath the sea. The areas around subsurface hydrothermal vents are physiologically more productive than the rest of the deep sea, frequently housing complex ecosystems driven by the chemicals dissolved in the vent fluids. Chemosynthetic bacteria and archaea are the foundation of the food chain, supporting a wide range of organisms such as giant tube worms, clams, limpets, and shrimp. Active hydrothermal vents are thought to exist on Jupiter's moon Europa, and Saturn's moon Enceladus, and it is speculated that ancient hydrothermal vents once existed on Mars.

Deep ocean hydrothermal vents frequently arise along mid-ocean ridges (MORBs) such as the East Pacific Rise and the Mid-Atlantic Ridge. These are areas where two tectonic plates are separating and new crust is forming.

The water emitted by seafloor hydrothermal vents is primarily seawater brought into the hydrothermal system near to the volcanic edifice via faults and porous sediments or volcanic layers, with some magmatic water released by upwelling magma.

The majority of water circulated within the fumarole and geyser systems in terrestrial hydrothermal systems is meteoric water plus groundwater that has percolated down into the thermal system from the surface, but it also commonly contains some portion of metamorphic water, magmatic water, and sedimentary formational brine that is released by the magma. The proportion of each varies depending on region (Müller et al., 2019).

A volcano is a crack in the crust of a planetary-mass object like Earth that permits hot lava, volcanic ash, and gases to escape from a magma chamber under the surface. It is a kind of vent.

Volcanic fissure vents are horizontal, linear fractures through which lava flows. Shield volcanoes are generated by the eruption of low-viscosity lava that can flow a long distance from a vent.

Slow eruptions of exceedingly viscous lava form lava domes. They can form within the crater of a previous volcanic eruption, as was the situation with Mount St. Helens, or they can form independently, as was the case with Lassen Peak. They, like stratovolcanoes, can erupt violently and explosively, but the lava usually does not flow far from the erupting vent.

Cinder cones are formed by the accumulation of minute fragments of scoria and pyroclastics (both of which resemble cinders, hence the name of this volcanic type). These can be short-lived eruptions that form a cone-shaped hill 30 to 400 meters (100 to 1,300 feet) high. The majority of cinder cones explode only once. Cinder cones can emerge as flank eruptions on bigger volcanoes or as stand-alone structures.

Visibility

Visibility is a meteorological term that refers to the distance at which an object or light may be seen clearly. It is determined only by the transparency of the surrounding air, and as such, it remains constant regardless of the ambient light level or time of day.

Depending on the country, it is recorded within surface weather observations and METAR code in meters or statute miles. Visibility influences all modes of transportation, including automobiles, sailing, and flying (Metz, 2008).

Several physical and optical aspects influence visibility perception. A realistic definition should take into account the fact that the human visual system (HVS) is highly sensitive to spatial frequencies, and then assess visibility using the Fourier transform and the HVS's contrast sensitivity function.

Visibility less than 100 meters (330 feet) is commonly reported as zero. Roads may be closed in certain situations, or automatic warning lights and signs may be activated to warn traffic. These have been installed in regions where visibility is consistently poor, such as after traffic collisions or pile-ups involving many vehicles.

Furthermore, a government meteorological organization, such as the United States National Weather Service, will frequently issue an advisory for limited visibility, such as a dense fog advisory. These typically warn drivers to avoid driving until the fog clears or other conditions improve. Low visibility also frequently delays airport traffic, resulting in long waits due to approach visibility minimums and the difficulties of safely relocating planes on the ground in low visibility.

Visibility in science refers to whether or not an object can be seen clearly, as well as the distance from which it can be observed. This is especially true when discussing weather and astronomy. A variety of factors, including weather, pollution, and time of day, can all have an impact on visibility. These factors can modify the state of the atmosphere, impacting the clarity of the air and the ability to see objects, particularly those that are far away.

Volcano

A volcano is a crack in the crust of a planetary-mass object like Earth that permits hot lava, volcanic ash, and gases to escape from a magma chamber under the surface.

Volcanoes are most commonly located where tectonic plates are diverging or converging on Earth, and the majority are found underwater. The Mid-Atlantic Ridge, for example, features volcanoes created by divergent tectonic plates, whereas the Pacific Ring of Fire has volcanoes caused by converging tectonic plates (Li and Li, 2016). Volcanoes can also form when the crust's plates are extending and thinning, such as in the East African Rift, the Wells Gray-Clearwater volcanic zone, and the Rio Grande Rift in North America. Volcanism distant from plate boundaries is thought to be caused by upwelling diapirs from the core-mantle barrier, which is located 3,000 kilometers (1,900 miles) below the Earth's surface.

As ash and droplets of sulfuric acid hide the Sun and cool the Earth's troposphere, large eruptions can alter atmospheric temperature. Large volcanic eruptions have historically been followed by volcanic winters, which have resulted in devastating famines (Figure 170).

Figure 170. Fountain of lava erupting from a volcanic cone in Hawaii, 1983.

Source: https://en.wikipedia.org/wiki/Volcano#/media/File:Lava_fountain_USGS_page_30424305-068_large.JPG.

Volcanoes are most commonly located where tectonic plates are diverging or converging on Earth, and the majority are found underwater. The Mid-Atlantic Ridge, for example, features volcanoes created by divergent tectonic plates, whereas the Pacific Ring of Fire has volcanoes caused by converging tectonic plates.

Volcanoes can also form when the crust's plates are extending and thinning, such as in the East African Rift, the Wells Gray-Clearwater volcanic zone, and the Rio Grande Rift in North America. Volcanism distant from plate boundaries is thought to be caused by upwelling diapirs from the core-mantle barrier, which is located 3,000 kilometers (1,900 miles) below the Earth's surface (Ledley et al., 2011). This leads in hotspot volcanism, such as the Hawaiian hotspot. Volcanoes are rarely formed where two tectonic plates glide past each other. The term volcano is derived from the name Vulcano, a volcanic island in Italy's Aeolian Islands, whose name is derived from Vulcan, the Roman god of fire. Volcanology, also spelled vulcanology, is the study of volcanoes.

Volume

The amount of three-dimensional space enclosed by a closed surface is expressed as a scalar quantity. For example, the space occupied or contained by a substance (solid, liquid, gas, or plasma) or 3D shape. Volume is frequently mathematically quantified using the SI-derived unit, the cubic meter. The volume of a container is commonly believed to be its capacity; that is, the amount of fluid (gas or liquid) that the container can hold, rather than the amount of space that the container occupies. Volumes are also ascribed to three-dimensional mathematical shapes. Arithmetic formulas can quickly compute the volumes of various simple shapes, such as regular, straight-edged, and circular shapes. If a formula for the shape's boundary exists, volumes of complicated shapes can be calculated using integral calculus. In three-dimensional space, one-dimensional figures (such as lines) and two-dimensional forms (such as squares) have zero volume.

Fluid displacement can be used to calculate the volume of a solid (whether it is regular or irregular in shape). Liquid displacement can also be used to calculate the volume of a gas. The combined volume of two substances is usually bigger than the volume of either substance alone. However, sometimes one material dissolves in the other, and the combined volume is not additive in such circumstances. A substance's volume is the amount of space it takes up, whereas its mass is the amount of stuff it contains. Density is defined as the quantity of mass per unit volume. Volume without mass is possible, as in an enclosed vacuum (Lewis, 2008).

The three-dimensional space occupied by materials or surrounded by a surface is referred to as volume. The cubic meter (m^3) is the SI unit of volume, however there are numerous others. Other volume units include the liter (L), cubic centimeter (cm^3), fluid ounce (fl oz), pint (pt), quart (qt), gallon (gal), teaspoon (tsp), and tablespoon (tbls).

The volume of a system is a key comprehensive parameter in thermodynamics for describing its thermodynamic state. The specific volume, an intense attribute, is the volume of the system per unit mass. Volume is a function of state and is related to other thermodynamic characteristics like pressure and temperature. The ideal gas law, for example, relates volume to pressure and temperature of an ideal gas. A system's physical volume may or may not correspond to the control volume used to study the system.

W

Wastage

Mass wasting, also known as mass movement, refers to the movement of rock or soil down slopes due to gravity. The debris transported by mass wasting is not entrained in a flowing medium, such as water, wind, or ice, as is the case with other erosion processes. Creep, solifluction, rockfalls, debris flows, and landslides are all examples of mass wasting, each with its own distinct characteristics and occurring over periods ranging from seconds to hundreds of years. Mass waste has been seen on Earth, Mars, Venus, Jupiter's moons Io, and many other Solar System entities (Cerase, 2019).

Subsidence is sometimes considered a type of mass waste. Then, a distinction is drawn between mass wasting caused by subsidence, which involves little horizontal movement, and mass wasting caused by slope movement. Landslides, for example, can be fatal and catastrophic due to their rapid mass waste. Soil creep, a slower mass loss, provides issues to civil engineering because it can deform roadways and structures and break pipelines. Slope stabilization, the construction of walls, catchment dams, or other structures to confine rockfall or debris flows, afforestation, or enhanced drainage of source regions are all means of mitigation.

The phrase "mass wasting" refers to any type of erosion that is caused by gravity and in which the carried soil and rock is not entrained in a flowing media such as water, wind, or ice. Water normally facilitates mass waste, although it is not abundant enough to be considered a carrying medium. Thus, without a sharp dividing line, the distinction between mass wasting and stream erosion lies between a mudflow (mass wasting) and a highly muddy stream (stream erosion) (Figure 171) (Košler, 2007).

Figure 171. A 300-meter-long slump that occurred in an area of thawing permafrost.

Source: https://www.nps.gov/subjects/erosion/mass-wasting.htm.

There are numerous types of mass wasting, each with its own distinct characteristics and occurring across durations ranging from seconds to hundreds of years.

Mass movements are classed as creeps or landslides based on how the soil, regolith, or rock flows downslope as a whole. Subsidence is sometimes considered a type of mass waste. Then, a distinction is drawn between mass wasting caused by subsidence, which involves little horizontal movement, and mass wasting caused by slope movement (Cossu et al., 2010).

Mass wasting occurs on both terrestrial and underwater slopes on Earth. Submarine mass wasting is most widespread around glaciated coastlines, where glaciers melt and large amounts of silt are released. Submarine slides may transport massive amounts of silt over hundreds of kilometers in a matter of hours.

Mass wasting is a typical phenomenon that occurs across the Solar System when volatile minerals are lost from regolith. This type of mass waste has been observed on Mars, Io, Triton, and possibly Europa and Ganymede. Mass wasting also occurs in the equatorial areas of Mars, where wind erosion steepens stopes of soft sulfate-rich sediments. The rough topography of tesserae is related with mass loss on Venus. Io's volcanic mountains have experienced substantial mass wasting.

Water Level

The water table is the zone of saturation's upper surface. The zone of saturation is the area of the ground where the pores and fractures are saturated with water. It can alternatively be defined simply as the depth below which the Earth becomes saturated.

The water table is the point at which the water pressure heads equal air pressure (gauge pressure = 0). It can be thought of as the "surface" of subsurface materials that are saturated with groundwater in a specific area.

Groundwater can come from either precipitation or groundwater seeping into the aquifer. Water infiltrates through pore spaces in the soil, passing through the unsaturated zone, in places with sufficient precipitation. Water fills more of the pore spaces of soils as depth increases, until a saturation zone is achieved.

Aquifers are layers of permeable rock that provide groundwater beneath the water table, in the phreatic zone (zone of saturation). The water table may be more difficult to determine in less permeable soils, such as tight bedrock formations and past lakebed deposits(Clary and Wandersee, 2011).

The water table is not the same as the water level in a deeper well. If a deeper aquifer has a lower permeability unit that restricts upward flow, the water level in this aquifer may rise to a level greater or less than the elevation of the real water table. The pressure in the deeper aquifer determines the height of the water in this deeper well, which is referred to as the potentiometric surface rather than the water table.

Water Table

The water table is an underground border that separates the soil surface from the area where groundwater saturates crevices between sediments and rock fissures. At this point, the pressures of water and air are equal.

The unsaturated zone is the soil surface above the water table, where both oxygen and water fill the crevices between particles. Because of the presence of oxygen in the soil, the unsaturated zone is also known as the aeration zone. The saturated zone is located beneath the water table, where water fills all crevices between sediments. The saturated zone is surrounded on all sides by impermeable rock.

The water table's shape and height are determined by the land surface above it; it curves up behind hills and lowers beneath valleys. The groundwater found under the water table is the result of rainwater that has seeped through surface soil (Clowes, 2010). Springs arise where the water table meets the land surface naturally, causing groundwater to flow from the surface and eventually into a stream, river, or lake.

The water table level can vary from one place to the next, and even within the same area. Changes in precipitation across seasons and years produce fluctuations in the water table level. The water table rises in late winter and early spring as snow melts and precipitation increases. However, there is a time lag between when precipitation enters the saturated zone and when the water table rises. This is due to the fact that water takes time to trickle through the crevices between sediments to reach the saturated zone, despite the fact that gravity aids the process. Crop irrigation can also raise the water table by allowing extra water to soak into the ground (Figure 172).

Figure 172. A freshwater spring in the desert of Iran. These springs show the importance of water tables in sustaining life in the harshest parts of Earth.

Source: https://www.nationalgeographic.org/encyclopedia/water-table/.

The water table tends to decline throughout the summer months, due in part to plants sucking up water from the soil surface before it can reach the water table. Human extraction of groundwater using wells also has an impact on the water table level; groundwater is pumped out for drinking water and to irrigate fields (Kleinhans et al., 2010). The depth of the water table in existing wells can be monitored to identify the impacts of season, climate, or human impact on groundwater. Water table measurements gathered from wells can be used to map the water table across regions.

The water table may permanently decline if water is not removed from a well in a sustainable manner. This is becoming more common all around the world. Some of the world's most important groundwater sources are being depleted to the point where they can no longer be refilled in India, China, and the United States. Groundwater depletion occurs when the rate of groundwater extraction through wells is higher than the rate of replenishment from precipitation.

Wave

A wave is the regular and structured transmission of disturbances from one location to another. Surface waves on water are the most well-known example, but sound, light, and the motion of subatomic particles also exhibit wave-like qualities. The disturbance oscillates periodically (see periodic motion) with a set frequency and wavelength in the simplest waves. Mechanical waves, such as sound, require a medium to flow through, whereas electromagnetic waves (see electromagnetic radiation) do not and can move in a vacuum. The propagation of a wave across a medium is determined by the qualities of the medium.

Waves are classified into two types: longitudinal and transverse. Transverse waves are similar to those found on water, with the surface moving up and down, but longitudinal waves are similar to those found in sound, with alternating compressions and rarefactions in a medium. The high point of a transverse wave is referred to as the crest, and the low point is referred to as the trough (Klippel et al., 2019). The compressions and rarefactions of longitudinal waves are identical to the crests and troughs of transverse waves. The wavelength is defined as the distance between successive crests or troughs. The amplitude of a wave is its height. The frequency is defined as the number of crests or troughs that pass through a certain spot in a unit of time. The wavelength multiplied by the frequency equals the velocity of a wave.

Even if the oscillation at one place is relatively slight, waves can travel enormous distances. A thunderclap, for example, can be heard from many kilometers away, but the sound transported reveals itself at each place only as minute compressions and rarefactions of the air (Figure 173).

Figure 173. Lunging breaker.

Source: https://en.wikipedia.org/wiki/Breaking_wave.

Wavelength

The wavelength of a periodic wave is defined in physics as its spatial period—the distance over which the wave's structure repeats. It is the distance between two successive corresponding locations of the same phase on the wave, such as two neighboring crests, troughs, or zero-crossings, and it is a feature of both traveling and standing waves, as well as other spatial wave patterns (Hundman and Mattmann, 2017). The spatial frequency is the inverse of the wavelength. The Greek letter lambda () is widely used to represent wavelength. The term wavelength is also used to describe modulated waves, as well as the sinusoidal envelopes of modulated waves or waves created by the interference of several sinusoids.

In the case of a sinusoidal wave moving at a constant speed, wavelength is inversely proportional to frequency: waves with higher frequencies have shorter wavelengths, whereas waves with lower frequencies have longer wavelengths.

The wavelength of a wave is determined by the medium through which it travels (for example, vacuum, air, or water). Sound waves, light waves, water waves, and periodic electrical signals in a conductor are all examples of waves. A sound wave is a variation in air pressure, whereas the intensity of the electric and magnetic fields varies in light and other electromagnetic radiation. Water waves are fluctuations in a body of water's height. Atomic locations change in a crystal lattice vibration (Figure 174).

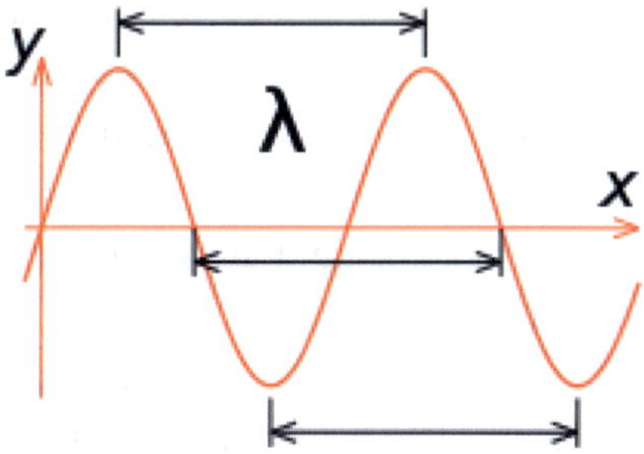

Figure 174. The wavelength of a sine wave, λ, can be measured between any two points with the same phase, such as between crests (on top), or troughs (on bottom), or corresponding zero crossings as shown.

Source: https://en.wikipedia.org/wiki/Wavelength.

A spectrum is the range of wavelengths or frequencies for wave phenomena. The term started with the visible light spectrum, but it is now used to refer to the full electromagnetic spectrum, as well as a sound or vibration spectrum.

Weather Wetland

A wetland is a separate habitat that is permanently (for years or decades) or seasonally flooded by water (for weeks or months). Flooding causes oxygen-free (anoxic) processes to predominate, particularly in soils. The peculiar flora of aquatic plants, adapted to the unique anoxic hydric soils, is the primary component that distinguishes wetlands from terrestrial land formations or water bodies (Guo et al., 2016). Wetlands are among the most biologically diverse habitats, supporting a broad variety of distinct plant and animal species. Many locations across the world have developed methods for monitoring wetland functions, ecological health, and overall wetland status. These strategies have helped to conserve wetlands by enhancing public knowledge of the functions that some wetlands provide.

Except for Antarctica, wetlands occur naturally on all continents. Wetlands contain either freshwater, brackish water, or saltwater. The primary wetland kinds are categorized based on the prevalent flora and/ or the water source. Marshes, for example, are wetlands dominated by emergent vegetation like reeds, cattails, and sedges, whereas swamps are wetlands dominated by woody vegetation like trees and bushes (although reed swamps in Europe are dominated by reeds, not trees). Tidal wetlands (ocean tides), estuaries (mixed tidal and river waters), floodplains (excess water from spilled rivers or lakes), springs, seeps, and fens (groundwater release out onto the surface), bogs, and vernal ponds are examples of wetlands classified by their water sources (rainfall or meltwater). Some wetlands feature many plant species and are nourished by multiple sources of water, making classification challenging.

Wetlands perform a variety of services that benefit people. These are known as ecosystem services, and they include water purification, groundwater replenishment, shoreline stabilization and storm protection, water storage and flood control, carbon processing (carbon fixation, decomposition, and sequestration), other nutrients and pollutants, and plant and animal support (Goldstein et al., 2017).

Wetlands serve as biodiversity reservoirs and provide wetland products. They also play a role in the mitigation and adaptation to climate change. However, some wetlands are substantial sources of methane emissions, and some are also nitrous oxide emitters. Wetlands are constructed to treat municipal and industrial wastewater as well as divert stormwater runoff. Wetlands that have been built may also play a part in water-sensitive urban planning (Figure 175).

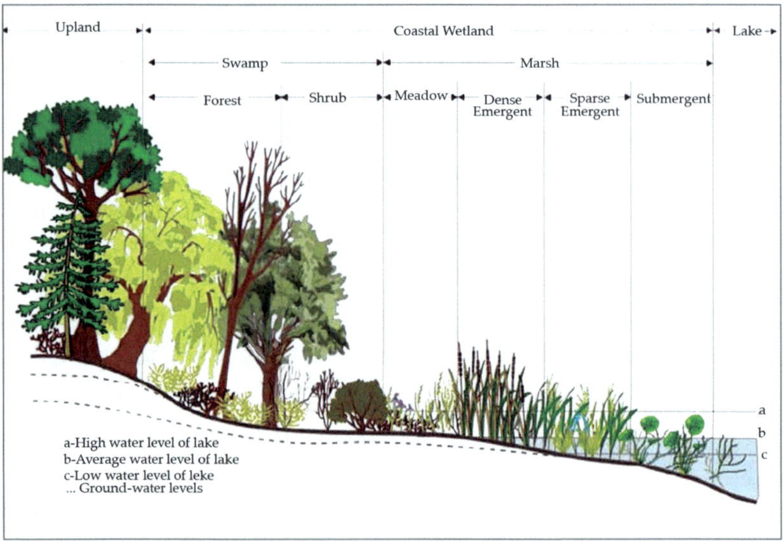

Figure 175. Upland vs. wetland vs. lacustrine zones.

Source: https://en.wikipedia.org/wiki/Wetland#/media/File:USGS_image_cropped.jpg.

Wind

Wind is the natural movement of air or other gases in relation to the surface of a planet. Wind occurs on a variety of scales, from thunderstorm flows lasting tens of minutes to local breezes formed by heating of land surfaces and lasting a few hours to global winds caused by differences in solar energy absorption between Earth's temperate zones (Escorihuela, 2018; Duffy, 2014). The differential heating between the equator and the poles, as well as the planet's rotation, are the two primary sources of large-scale atmospheric circulation (Coriolis effect). Thermal low circulations over topography and high plateaus in the tropics and subtropics can trigger monsoon circulations. Local winds in coastal locations can be defined by the sea breeze/land breeze cycle; in areas with diverse terrain, mountain, and valley breezes can prevail.

Long-duration winds are known by a variety of titles based on their average strength, including breeze, gale, storm, and hurricane (Hundman and Mattmann, 2017). Solar wind is the passage of gases or charged particles from the Sun across space, whereas planetary wind is the outgassing of light chemical components into space from a planet's atmosphere. The strongest winds yet detected on a planet in the Solar System are found on Neptune and Saturn.

BIBLIOGRAPHY

1. Bohle, M., (2015). *Simple Geoethics: An Essay on Daily Earth Science, 419*(1), 5–12. Geological Society, London, Special Publications.
2. Britner, S. L., (2008). Motivation is high school science students: A comparison of gender differences in life, physical, and Earth science classes. *Journal of Research in Science Teaching: The Official Journal of the National Association for Research in Science Teaching, 45*(8), 955–970.
3. Butler, D., (2016). Europe's GALILEO satellites herald new era for Earth science. *Nature News, 540*(7634), 493.
4. Cerase, A., (2019). Geoscientists' voice in the media: Framing Earth science in the aftermath of Emilia 2012 and Amatrice 2016 seismic crises. *Earthquake Risk Perception, Communication and Mitigation Strategies Across Europe, Il Sileno Edizioni, Geographies of the Anthropocene Book Series, Rende, 2*(2), 123–144.
5. Clary, R. M., & Wandersee, J. H., (2011). A "coprolitic vision" for Earth science education. *School Science and Mathematics, 111*(6), 262–273.
6. Clowes, R. M., (2010). Initiation, development, and benefits of lithoprobe—shaping the direction of Earth science research in Canada and beyond. *Canadian Journal of Earth Sciences, 47*(4), 291–314.
7. Cossu, R., Petitdidier, M., Linford, J., Badoux, V., Fusco, L., Gotab, B., & Vetois, G., (2010). A roadmap for a dedicated Earth science grid platform. *Earth Science Informatics, 3*(3), 135–148.
8. Duffy, T., (2014). Earth science: Crystallography's journey to the deep Earth. *Nature News, 506*(7489), 427.

9. Escorihuela, J., (2018). The role of the geotouristic guide in Earth science education: Towards a more critical society of land management. *Geoheritage, 10*(2), 301–310.

10. Goldstein, J. C., Mayernik, M. S., & Ramapriyan, H. K., (2017). Identifiers for Earth science data sets: Where we have been and where we need to go. *Data Science Journal, 16*.

11. Guo, H., Wang, L., & Liang, D., (2016). Big Earth data from space: A new engine for Earth science. *Science Bulletin, 61*(7), 505–513.

12. Hundman, K., & Mattmann, C. A., (2017). *Measurement Context Extraction from Text: Discovering Opportunities and Gaps in Earth Science.* arXiv preprint arXiv:1710.04312.

13. Kleinhans, M. G., Buskes, C. J., & De Regt, H. W., (2010). Philosophy of Earth science. *Philosophies of the Sciences: A Guide,* 289–317.

14. Klippel, A., Zhao, J., Jackson, K. L., La Femina, P., Stubbs, C., Wetzel, R., & Oprean, D., (2019). Transforming Earth science education through immersive experiences: Delivering on a long-held promise. *Journal of Educational Computing Research, 57*(7), 1745–1771.

15. Košler, J., (2007). Laser ablation ICP—MS—A new dating tool in Earth science. *Proceedings of the Geologists' Association, 118*(1), 19–24.

16. Ledley, T. S., Dahlman, L., McAuliffe, C., Haddad, N., Taber, M. R., Domenico, B., & Grogan, M., (2011). Making Earth science data accessible and usable in education. *Science, 333*(6051), 1838, 1839.

17. Lewis, E. B., (2008). Content is not enough: A history of secondary Earth science teacher preparation with recommendations for today. *Journal of Geoscience Education, 56*(5), 445–455.

18. Li, X. H., & Li, Q. L., (2016). Major advances in microbeam analytical techniques and their applications in Earth science. *Science Bulletin, 61*(23), 1785–1787.

19. Liou, Y. A. (2015). Seminar@ IHOS :(Sep/10) Prof. Yuei-An Liou from Center for Space and Remote Sensing Research, NCU (14: 00, Room S-325, Earth Science Building# 1).

20. Metz, S., (2008). Earth in the balance. *The Science Teacher, 75*(1), 8.

21. Movva, S., Ramachandran, R., Maskey, M., Kulkarni, A., Conover, H., & Nair, U. S., (2012). Talkoot: Software tool to create collaboratories for Earth science. *Earth Science Informatics, 5*(1).

22. Müller, J., Murphy, T. W., Schreiber, U., Shelus, P. J., Torre, J. M.,

Williams, J. G., & Hofmann, F., (2019). Lunar laser ranging: A tool for general relativity, lunar geophysics and Earth science. *Journal of Geodesy, 93*(11), 2195–2210.

23. Neeck, S. P., & Volz, S. M., (2011). NASA's Earth science flight program overview. In: *Sensors, Systems, and Next-Generation Satellites XV* (Vol. 8176, p. 817602). International Society for Optics and Photonics.
24. Odenwald, S. F., (2020). A citation study of Earth science projects in citizen science. *Plos One, 15*(7), e0235265.
25. Oreskes, N., (2013). Earth science: How plate tectonics clicked. *Nature News, 501*(7465), 27.
26. Oreskes, N., (2015). How Earth science has become a social science. *Historical Social Research/Historische Sozialforschung,* 246–270.
27. Orion, N., (2019). The future challenge of Earth science education research. *Disciplinary and Interdisciplinary Science Education Research, 1*(1), 1–8.
28. Özyurt, Ö., Cagiltay, N. E., Özyurt, H., & Akgun, A., (2021). A systematic review and mapping of the literature of virtual reality studies in Earth science engineering education. *Journal of Pedagogical Research, 5*(2), 237–256.
29. Parkes, M. A., & Morris, J. H., (2001). Earth science conservation in Ireland: The Irish geological heritage program. *Irish Journal of Earth Sciences,* 79–90.
30. Pennington, D., Ebert-Uphoff, I., Freed, N., Martin, J., & Pierce, S. A., (2020). Bridging sustainability science, Earth science, and data science through interdisciplinary education. *Sustainability Science, 15*(2), 647–661.
31. Perdigão, R. A., Ehret, U., Knuth, K. H., & Wang, J., (2020). Debates: Does information theory provide a new paradigm for Earth science? Emerging concepts and pathways of information physics. *Water Resources Research, 56*(2), e2019WR025270.
32. Plotnick, R. E., Stigall, A. L., & Stefanescu, I., (2014). Evolution of paleontology: Long-term gender trends in an earth-science discipline. *GSA Today, 24*(11), 44, 45.
33. Press, F., (2008). Earth science and society. *Nature, 451*(7176), 301–303.
34. Pyle, E. J., Darling, A., Kreager, Z., & Conrad, S. H., (2018). Research on students' conceptual understanding of geology/solid Earth science

content. *A Community Framework for Geoscience Education Research*, 9.

35. Rao, Y. B., (2012). Interaction meet on contribution to Earth science research over the last fifty years. *Journal of Geological Society of India* (Online archive from Vol. 1–78), *79*(1), 103–109.
36. Rivet, A. E., & Kastens, K. A., (2012). Developing a construct-based assessment to examine students' analogical reasoning around physical models in Earth Science. *Journal of Research in Science Teaching, 49*(6), 713–743.
37. Schmitz, B., (2011). Earth science: Mind your head. *Nature, 471*(7340), 573–574.
38. Schmitz, B., (2012). Earth science: How a world came to be. *Nature, 485*(7396), 39–39.
39. Showstack, R., (2008). European Earth science network for postdocs. *EOS, Transactions American Geophysical Union, 89*(49), 503–503.
40. Telling, J., Lyda, A., Hartzell, P., & Glennie, C., (2017). Review of Earth science research using terrestrial laser scanning. *Earth-Science Reviews, 169*, 35–68.
41. Thessen, A., (2016). Adoption of machine learning techniques in ecology and Earth science. *One Ecosystem, 1*, e8621.
42. Tollefson, J., (2015). Earth science wrestles with conflict-of-interest policies. *Nature News, 522*(7557), 403.
43. Toth, C., Pethe, M., & Hathazi, A., (2014). The application of Earth science-based analyses on a twin-kurgan in Northern Hungary. *Carpathian Journal of Earth and Environmental Sciences, 9*(1), 11–20.
44. Wallace-Wells, D., (2019). *The Uninhabitable Earth* (pp. 271–294). Columbia University Press.
45. Wilkinson, C., Hikuroa, D. C., Macfarlane, A. H., & Hughes, M. W., (2020). Mātauranga Māori in geomorphology: Existing frameworks, case studies, and recommendations for incorporating Indigenous knowledge in Earth science. *Earth Surface Dynamics, 8*(3), 595–618.
46. Witze, A., (2013). Earth science: Under the volcano. *Nature News, 504*(7479), 206.
47. Yu, L., & Gong, P., (2012). Google Earth as a virtual globe tool for Earth science applications at the global scale: Progress and perspectives. *International Journal of Remote Sensing, 33*(12), 3966–3986.

48. Zhuang, J., Jacob, D. J., Gaya, J. F., Yantosca, R. M., Lundgren, E. W., Sulprizio, M. P., & Eastham, S. D., (2019). Enabling immediate access to Earth science models through cloud computing: Application to the GEOS-chem model. *Bulletin of the American Meteorological Society, 100*(10), 1943–1960.

INDEX

L

M

N

O

P

Q

R

S

Printed in the United States
by Baker & Taylor Publisher Services